The relations between the sciences

THE RELATIONS BETWEEN THE SCIENCES

by the late

C. F. A. PANTIN

Professor of Zoology
in the University of Cambridge, 1959–1966

Edited and with an Introduction
and Notes by

A. M. PANTIN *and* W. H. THORPE

Based upon the Tarner Lectures
of Trinity College, 1959

CAMBRIDGE
AT THE UNIVERSITY PRESS
1968

Published by the Syndics of the Cambridge University Press
Bentley House, 200 Euston Road, London, N.W. 1
American Branch: 32 East 57th Street, New York, N.Y. 10022

Library of Congress Catalogue Card Number: 68-30952
Standard Book Number: 521 05909 7

Printed in Great Britain
at the University Printing House, Cambridge
(Brooke Crutchley, University Printer)

Preface

CARL PANTIN was a man of exceptional breadth and depth of learning. By profession he was a zoologist, but the whole natural world inspired him with delight and wonder, and he had that curiosity of mind which is a rich gift to the possessor. This tireless interest in objects and events around him gave Carl, besides a great deal else, considerable knowledge of certain sciences other than that in which his main work lay.

In zoology, his knowledge and understanding of the Invertebrata were unrivalled. His chief and pre-eminent contributions to knowledge in this field were his electro-physiological and histological studies of the simpler types of nervous system, especially of the nerve net of the sea-anemone and the nervous system of some Crustacea. It was these studies that led to his major scientific honours, such as the Royal Medal of the Royal Society and the Croonian Lectureship. The subject of that lecture was 'The Elementary Nervous System'. These studies of the mode of action of this very simple and primitive apparatus, which is nevertheless capable of organising and controlling behavioural reactions of surprising complexity, were the seed from which developed a strong interest in ethology.

'Scratch a biologist and you find a philosopher' was a remark which Carl liked to quote. He was gratified to be invited to give the Tarner Lectures at Trinity College, and chose as his subject 'The Relations between the Sciences', which is in fact almost identical with the terms of the Tarner Foundation. As a biologist, and one with wide acquaintance with other sciences and especially with physics and geology, Carl was outstandingly qualified to lecture on this theme. Other qualifications he had which will be apparent to his readers.

The lectures were delivered in 1959, shortly after Carl had been appointed to the Chair of Zoology at Cambridge. As Professor he at once devoted himself with energy to the innumerable tasks involved in reorganising a large teaching and research department, tasks which included the planning and erection of a new museum and research building. While carrying this heavy burden of teaching and administration he contrived to pursue some of his own research. All this and many other duties he undertook with that gay and generous enthusiasm before which difficulties gradually melted away and which secured the whole-hearted co-operation, one can truthfully say the love, of colleagues of widely different temperaments and interests. He continually hoped soon to have enough time and peace of mind to prepare the 'Tarners' for press. Each year he would hopefully say, 'This long vacation will see the end of the job'. But it was, alas, not to be, and the task was still unfinished when he died in January 1967.

When we came to consider the problem of publication we were delighted to find that all the lectures had been thoroughly revised and the MS was in a more advanced state than we had dared to hope. It was clear from pencilled notes and comments, and also from a good deal of discussion with him in the last year or two, that Carl had intended a further revision of at least some parts. Nevertheless, we feel confident that the version left us was not far from the definitive state.

Accordingly, the Editors' task has been a light one. The illustrations had all to be chosen, the references completed and corrected, summaries of three of the chapters written and the index made. Then there was the problem of notes. In most cases the pencilled marginalia were not clear enough to warrant any attempt at modifying passages in the text. But they did, together with other lines of evidence, indicate what problems were particularly in Carl's mind in recent years and the direction in which his thought was moving. The notes have therefore been planned with the idea of drawing attention to very recent books and papers which we felt it likely that Carl would have wished to discuss or at least carefully to consider in any further revision he might have done. Besides these notes, which we hope readers both specialist and general will find helpful, we have reproduced as appendices three papers of Carl's, written since he gave the Tarner Lectures, which carry his ideas on certain important topics some steps further. It seemed best to reprint these exactly as Carl wrote them and not to attempt to avoid occasional overlaps by making cuts or alterations. We are grateful to the publishers of these papers for allowing us to reprint them. We record our thanks to several friends who have helped us, especially in tracing references. They know that we remember them individually and very gratefully.

A. M. P.
W. H. T.

Cambridge
August 1967

Contents

The name of physical science, however, is often applied in a more or less restricted manner to those branches of science in which the phenomena considered are of the simplest and most abstract kind, excluding the consideration of the more complex phenomena, such as those observed in living beings.

J. Clerk Maxwell, 1912

Physics and chemistry will dominate biology only by becoming biology.

A. V. Hill, 1954

I

The restricted and the unrestricted sciences

AT THEIR FOUNDATION, the subject matter of the Tarner Lectures was defined as 'the Philosophy of the Sciences and the Relations, or Want of Relations, between the different Departments of Knowledge'. It is dangerous for a man of science to meddle with philosophical conclusions, even though his work inevitably brings their importance before him. He can hear the echo of that great naturalist John Ray in his Preface to *The Wisdom of God*:[1] 'I am sensible that this Tractate may likely incur the Censure of a superfluous piece, and myself the Blame of giving the Reader unnecessary Trouble, there having been so much so well written of this subject by the most learned Men of our Time...' All the same, Natural Knowledge is increasing so rapidly, so enormously, and in such apparently unrelated ways, that it can be useful for a man of science to discuss the matter, if but an amateur in philosophy, for he is in some position to review the relations or want of relations between the different departments of scientific knowledge as he sees them in practice.

What the practical scientist sees today is indeed a puzzle. However much we choose to classify the sciences together, the nuclear physicist at Harwell and the systematic entomologist identifying and classifying his insects at the British Museum of Natural History at least seem to be trying to do very different things, and their methods seem to differ so greatly that it is by no means obvious why we should lump them together as scientists and talk about their methods of investigation as *the* scientific method.

I shall therefore begin by discussing some views about the apparent relationship of the sciences and the goals of scientific investigation as they are seen today. That leads to discussion of the nature of scientific analysis and the extent to which this necessarily consists of the interpretation of larger-scale phenomena in terms of ever smaller and briefer events. In turn that leads to consideration of the different levels of complexity of natural phenomena and of the manner in which different rules seem to govern gross phenomena from those which hold at and below the molecular level. The nature of living systems is then considered and the special significance for them of this boundary between the gross and molecular levels of structure. This leads us to the scientific importance of determining the class to which a phenomenon belongs and the significance of classification in the sciences. It raises the question of how far such classification is a reflection of the 'real' world, and how far it is a reflection of our own mental machinery; and so we are led to reconsider the scientific method in the different sciences.

Let us begin with the term 'science'. Strictly this means all knowledge. By popular usage it has become more or less restricted to knowledge about objects in the natural world, to natural science. I say 'more or less' because since the seventeenth century it has also included the abstract and deductive science of mathematics. 'The Royal Society of London for improving Natural Knowledge', to give it its full title, arose from 'a designe of founding a Colledge for the promoting of Physico-Mathematicall, Experimental Learning'; and *Chambers's Encyclopaedia*[2] today says: 'Traditionally science has been subdivided according to differences of subject matter, hence the familiar names—mathematics, physics, chemistry, biology and its various branches.' Indeed, the Royal Society itself, forced to make a convenient and practical division in its publications, now gives us 'Series A: Mathematical and Physical Sciences; Series B: Biological Sciences'.

When considering the relationship of the various sciences it at first sight seems convenient to place them in a serial order. Following *Chambers's Encyclopaedia* we might start with mathematics, go on to physics and chemistry, and then pass to biochemistry, physiology and the other biological sciences. But an attempt to arrange the various sciences in such a linear series is unsatisfactory because it cannot be made to include them all or to display all their relationships. Such an attempt recalls the old linear classification of the animal, vegetable and mineral kingdoms into a single *Scala Naturae*. Within the sciences botany and zoology must anyway be placed side by side, so that at least a branching series is required. But even greater difficulties are raised by geology. In the editorial classification of the Royal Society this is arbitrarily placed in the 'B' publications, unless the problem concerns such matters as magnetism or crystallography—which are just as much a concern of the theory of the earth as palaeontology. The newcomer biophysics is in an equally unfortunate position. The relationships of the sciences can in fact only be displayed by a multidimensional network which is highly inconvenient for the administrative machinery of our minds, accustomed as it is to reduce understanding to linear arguments and dichotomous classifications capable of development by deductive logic. Any divisions we make in the range of natural phenomena are in part a matter of our convenience.

At any period the divisions are strongly affected by their historical development. Cambridge, unlike most universities, treats mineralogy and geology as separate sciences. This originated in the tangled politics of this university at the beginning of the nineteenth century, as a result of which an excellent natural historian, J. S. Henslow—later Charles Darwin's director of studies—became Professor of Mineralogy. Almost at once he switched to the Chair of Botany, a science which he preferred. But, the political heat having hatched it, the Chair of Mineralogy continued and on Henslow's resignation passed to William Whewell and on to the present day.

Our accepted divisions of the sciences owe much to the accident of eighteenth-century benefactions, to botany, to human anatomy, and so on. But this does not mean that our divisions are wholly arbitrary. Thus, however much we properly accent the unity of the biological sciences, plants on the one hand and animals on the other are distinct classes of living things, each with their own associated phenomena. The distinction between botany and zoology is not arbitrary. It can be traced to the fact that plants are essentially photosynthetic machines in which the essential adaptive machinery which gives character to the object is at the cellular level of the molecule chlorophyll, and of the minute cellular organs the chloroplasts; whereas animals are essentially predatory behaviour machines, the adaptive character of whose machinery is to be seen at a gross anatomical level of eyes, brains and muscles. Yet even here history has its effect. For bacteria and other micro-organisms constitute a third and different class of living objects which could not be appreciated until microscopes had reached the necessary state of perfection in the nineteenth century. Our classes of the sciences may be valid, but in any age we only appreciate those which fall within our 'sensory spectrum', aided as that is by contemporary instruments.

Of the various sciences, mathematics occupies a peculiar position. Today it has gained immense importance by supplying models to represent the recondite phenomena of the physicist, and at universities training in both these sciences is intimately linked. The mathematician and the physicist are more likely to understand each other's technical conversation than either would understand that of a museum taxonomist. Mathematics, the abstract science, thus seems closer to physics than does physics itself to another concrete science, taxonomy. Even though we must remember that the abstract processes of a mathematician may have a remarkable parallel with the physical operations of a calculating machine, and that the whole success of applied mathematics depends on the accuracy with which mathematical relations can be shown to correspond to what happens in the real world, it may still seem a little strange that abstract mathematics should seem so much nearer to the concrete science of physics than that is to the concrete science of taxonomy. As a science, mathematics is not only acknowledged, it is given pride of place: the Queen of the Sciences. And though one may still say,

> But which Pretender is and which is King
> God bless us all, that's quite another thing,

one has only to look into any scientific journal to see how mathematics ministers to every one of the concrete sciences.

But it is in physical science that the linkage with mathematics is most evident. Indeed, such is their joint prestige that the man in the street, faced with the tremendous impact of physical science and technology upon his everyday affairs, is apt to equate the whole of natural science with mathematics and physics—in London the 'Science Museum' is distinct from the 'Natural History Museum'. Prestige is of course to some extent a matter of fashion. At the beginning of the nineteenth century geology was dominant, and biology took that place after the publication of *The Origin of Species*. But the modern prestige of mathematical physics is more fundamental.

When we try to force the sciences into a linear series, however imperfectly, we are at least trying to exhibit a real quality in their relationship, one which has to do with mathematics. Evidently, in this series we are concerned with phenomena of increasing complexity. We may note that as we pass from biology to present-day physics we are passing from highly complex phenomena, which so often we are not clever enough to analyse, to simple ones which we can represent by mathematical models of exceedingly high intellectual penetration. The enormous advance of the physical sciences has in fact been rendered possible just because they are thus restricted in their scope. The more we restrict the class of phenomena we observe and the number of its variables, the more far-reaching are the possible deductive consequences of our hypotheses. But in so doing much of the grand variety of natural phenomena is systematically excluded from study. Very clever men are answering the relatively easy questions of the natural examination paper. In contrast, in biology such problems as 'What will be the ecological consequences of a general increase in nuclear radiation?' are so difficult that our answers seem paltry and emotional.

It is fair to say that at any period the different sciences are in different states of evolution. In medieval times Wisdom was divided into the three philosophies: Metaphysical, Moral, and

Natural philosophy covering what we would now call natural science. After the scientific revolution of the seventeenth century, Newton's enormous success in the application of mathematics to mechanics and astronomy gradually led to the restriction of the term natural philosophy to the physical sciences. By the beginning of the nineteenth century, however, chemistry had emerged as an experimental and exact science in its own right. Accordingly Playfair in his *Outlines of Natural Philosophy*[3] of 1812 divided the field of natural knowledge into (1) Natural Philosophy, which concerned action that takes place between bodies without permanent change in their internal constitution; (2) Chemistry, which concerned action that takes place between bodies producing permanent change in their internal constitution; (3) Natural History: 'The branch of knowledge which collects and classifies facts is called Natural History...its objects are confined to what are called the three kingdoms, Mineral, Vegetable and Animal.' Physical science and chemistry had achieved the status of 'exact' experimental sciences. The only parts of biology and geology to which reference is made remain at a simpler level of observation and description.

A good summary of the supposed position of the sciences at the beginning of this century is given by Mellone[4] in an elementary text-book of 1902:

> Without experiment, mechanics, physics and chemistry could scarcely exist; these are fundamental sciences in an advanced state. In physiology experiment plays a much smaller part, for if made at all it must be made on the organs of the living body. In the sciences of description and classification—botany, zoology and mineralogy—the range of experiment is more restricted; while in astronomy, geology and meteorology we may say that experiment so far as we are concerned is impossible. We say 'so far as we are concerned', because Nature sometimes produces phenomena of so remarkable a character that she may be said to be making an experiment herself—as in an 'eclipse of the sun'.

To this we may add the comment of Jevons[5] in 1870:

> Every question of science is first a matter of fact only, then a matter of quantity and by degrees becomes more and more precisely quantitative.

We thus get a picture of successive sciences becoming progressively more experimental and more quantitative. The experience of the last sixty years certainly seems to justify this. Mellone's summary, in fact, does less than justice to the experimental character of physiology even in 1902. Since then, along with mineralogy (particularly through crystallography), biochemistry and many others, it has become an exact science.

Consider the problem of nervous excitation, on which today so much of our interpretation of the action of the brain depends. Already before 1920 Keith Lucas[6] and Adrian[7] had demonstrated the nature of the impulse which passes along an excited nerve fibre. I shall discuss their work again later. For the moment I would remind you that their analysis depended on the application of brief electrical stimuli of measured intensity, duration and interval. By the use of such stimuli they showed that there was a precise threshold for stimulation of the nerve fibre which must be passed if a nervous impulse is to be propagated; and they showed that the propagated impulse was followed by a brief refractory period during which the nerve fibre was inexcitable. By their experimental analysis they showed that, provided the stimulus sufficed to provoke the impulse, the size of the impulse as measured by its effect was independent of the intensity of the stimulus. If the stimulus was effective at all, the effect was 'all or nothing'. What they did in these very important experiments was to establish the *class* of phenomenon to which the nervous impulse belonged: unlike the attenuating wave of sound that is sent down a speaking tube, the impulse maintains its intensity as it passes along a nerve fibre, just as a wave of combustion is maintained as it passes along a train of gunpowder. The establishment of the class to which a phenomenon belongs is precisely what a physicist did when he established the undulatory nature of sound, and what he is endeavouring to do with the properties of his mesons and hyperons. Classification is not just something done in museums.

Consideration of such experiments as those on nervous conduction suggests that the division between the exact and the descriptive sciences is moving back into the biological and geological sciences themselves. Not only do the individual sciences become increasingly quantitative and exact, but in so doing analysis of the phenomena with which each is concerned seems to become dependent on another science which deals with systems of smaller size and less complexity into which their parts can be broken up. The biological processes of the parts of an organism—nerves, muscles and so on—are analysed by the physiologist; the operation of these processes is then related to the biochemistry of its molecular changes, the interpretation of molecular combination is given by the chemist, and his work rests on what the physicist can tell us of the behaviour of atoms and electrons. The physicist himself during the last hundred years has carried the same process of analysis further. In the nineteenth century we had reached the atom, and that seemed to be the end of this sort of analysis. During the early years of this century atoms were decomposed into protons and electrons. Again for a brief period it looked as though we had reached finality. For general assumptions about the nature of the universe Eddington[8] in 1933 proposed that the number of particles in the universe could be enumerated. There were to be about 10^{79} electrons and a like number of protons. But this limit too was passed—and without any prediction that this would be the case. Today there are already known several dozen particles besides these: neutrons, mesons, hyperons and the rest; and to an outsider there seems no prospect of an end to this analysis, incomprehensible as the nature of the particles may be to him.

This tendency of sciences, when they become quantitative, to interpret their phenomena in terms of smaller grades of structure can be seen in biology itself. Anatomy at the beginning of the nineteenth century was concerned with the gross dissection of organs. During that century the development of the light microscope enabled analyses of organs into cells and finer structures. The power of analysis of the light microscope met a limit set by the wavelength of light. The invention of the electron microscope has opened up the analysis of cells into still finer structures. Does this imply that all the properties of complex structures are to be displayed by considering the properties of their simpler components, and so on? Are we to assume that, with the aid of the theory of probability, the mathematical physicist building the universe from his contemporaneously ultimate particles could, at least in theory, answer all our questions? One sees at once the difficulties in supposing this to be true of living organisms, but before we go on to biology let us consider the investigation of a problem among the complex structures of the inanimate world.

Geology deals with complex inanimate structures: rocks, rivers, mountains and glaciers. Let us consider the investigation of a problem of such things as these. One of the striking features of the scenery of this and other countries all over the world is the existence of high plains or plateaux. For instance, starting from the Welsh sea coast in Carmarthenshire, and walking up the valleys, we do not simply meet hills which get irregularly higher. At a height of about 500–600 feet we reach a plateau, a plateau dissected by rivers but stretching between and beyond them for a considerable distance. Farther inland there is evidence of a second plateau at a higher altitude. In these plains both the harder and the softer rocks are worn down to more or less the same level.

There is an extensive scientific literature about the origin of these plateaux.[9] Ramsay,[10] about a hundred years ago, ascribed them to marine denudation. Coastal erosion is a powerful agent and can undoubtedly plane down a land surface as it advances inland. At a much lower level round our coasts there are limited shelves and plains which can certainly be ascribed to this cause, for beach deposits of shells make it clear that they correspond to the oscillations of the level of the ocean during the Pleistocene. But the plateaux we are considering are higher and certainly much older.

The alternative to their marine origin is that these plateaux are peneplains; that is, they are the result of subaerial denudation acting on a land surface for an immense period, till the outstanding features have been worn away to a common level. Whichever view is taken, the total amount of rock which has been removed from the present land surface in Wales during the course of geological time is immense—a thickness of at least 10,000 feet of overlying strata has been removed.

There was no obvious way of deciding between the marine and the subaerial hypothesis for the origin of the plateaux. It is interesting to note that, as C. B. Travis pointed out in his review of the matter, geologists in Britain with their direct experience of coastal erosion were disposed towards the first view, whilst American geologists with their continental experience of subaerial

Figure 1. Upper part of the Towy Valley showing mature features (from O. T. Jones, 1924).

denudation in their Western Territory surveys held the second view. Like poetic imagery, scientific hypotheses have a historical background depending on our own experience. Neither Tennyson nor Meredith would have used imagery from the sea if they had spent their lives in the Middle West.

Important new evidence about the origin of the Carmarthenshire plateaux was produced some forty years ago by Professor O. T. Jones.[11] He had an intimate knowledge of the geology and geography of the plateaux, and was struck by certain features of their streams and valleys, particularly by the head waters of the drainage system of the upper Towy. He perceived that if the plateaux are the result of subaerial denudation the river systems might give some indication of this.

As a stream flows down from the hills we should expect that, other things being equal, the gradient of the stream would follow a regular decrease from its source down to the plains where it enters the sea. From their source the streams of the Towy system do indeed show a decrease of gradient; but there is a sudden steepening of slope between the upper and the lower parts of these valleys, a change which is not related to any difference in

Figure 2. The Towy above Fanog: rejuvenation beginning (from O. T. Jones, 1924).

the hardness of the rocks over which they run. In the upper part, the valleys are wide open with a gentle gradient and meadow-like flats on each side rising to gentle hills. But in the middle region of each valley there occurs a marked discontinuity. As we pass downstream, at a certain point the gradient rapidly steepens and the valleys plunge into V-shaped rocky gorges before leaving the plateau. The gradient then again gradually decreases and the valley opens out again as the river approaches the sea. There is every indication that the valleys are here being cut back or rejuvenated from the present level of the sea; whilst the upper parts of the streams represent an older state of the valleys before this rejuvenation had occurred.

By carefully surveying the profiles of the valleys it is possible to show that the gradual decreases of gradient along the course of a river can be represented by an empirical formula. In the lower part of the valley the application of the formula leads to an asymptotic convergence with sea level. But in the head waters above the point of discontinuity, application of the formula reveals that there the stream in a valley shows an asymptotic convergence

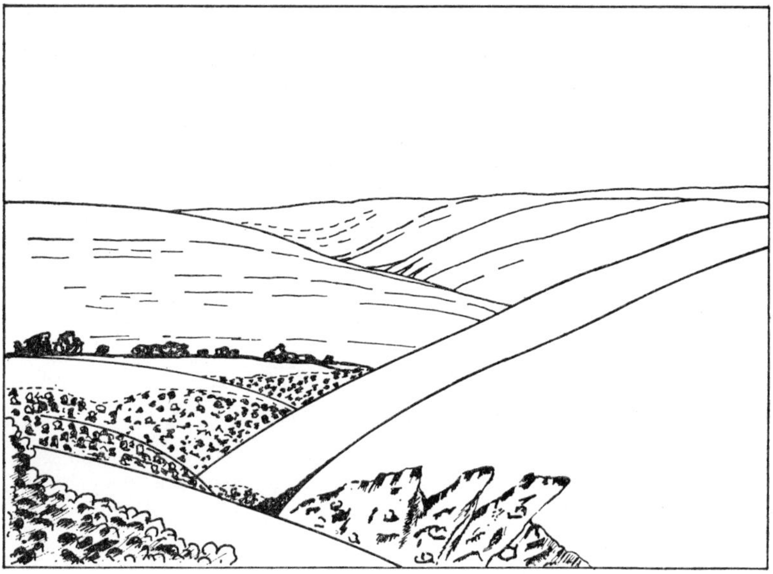

Figure 3. Trawsnant tributary valley showing rejuvenation of late mature valley (from O. T. Jones, 1924).

to a plain some 400 feet above the present sea level. Confirmation of this pattern and this base level is to be found in all the streams of the area. Thus, not only the plateaux themselves but the detailed character of the upper parts of their river systems are agreeable to the hypothesis that the plateaux are the result of prolonged subaerial denudation, but to a sea level 400 feet above that of the present day. In the discussion at the Geological Society on Professor Jones's paper it was suggested that the plateau was at least as old as the Pliocene. Some evidence of a higher and still older peneplain and river system was given in the paper.

This brief account necessarily leaves out many collateral issues which were discussed, such as the conclusion that the form of the valleys had not been seriously disturbed by glacial action during the Pleistocene. The report of the discussion on Professor Jones's paper shows not only a somewhat unusual unanimity in accepting a scientific speaker's argument but the immediate manner in which it was mentally applied by the audience to situations familiar to them in other parts of the world which they knew.

I myself first heard this work discussed long before I read the paper. That is, I learnt of the work on the authority of others. Nevertheless, I found the conclusions fully conformable with my own limited field of experience as an amateur geologist, and accepted them. I found it noteworthy that after learning of this work I immediately viewed various geological features that I encountered—for instance, in South Devon and in parts of Brazil—with a different eye. It is not that I have the same perceptions as before and then apply Professor Jones's reasoning to them; it is as though my actual mode of perception has been changed by the historical experience of this argument.

Of course, what I have here called my mode of perception may lead me into error. To give an example of this let us consider how a thorough knowledge of glaciers and glaciation and all their multitude of effects upon rocks, valleys, streams and deposits may both increase a man's perception and also mislead him. When Louis Agassiz visited Scotland in 1840, fresh from his experience of glacial action in Switzerland, he immediately recognised that almost every element of Scottish scenery bore witness to recent extensive glaciation. Today we can point and say at once, 'There is a moraine, there are roches moutonnées, there are glacial striae', as directly as we can say, 'There is a cigarette', or 'There is an apple'. We give our unquestioning assent to such statements. In general they are correct. When they are correct, our impressions of the natural world as they are presented to our minds have one character of supreme importance—their mutual consistency with every other impression we receive. If I 'rightly' perceive 'glacial striae' on a rock surface everything

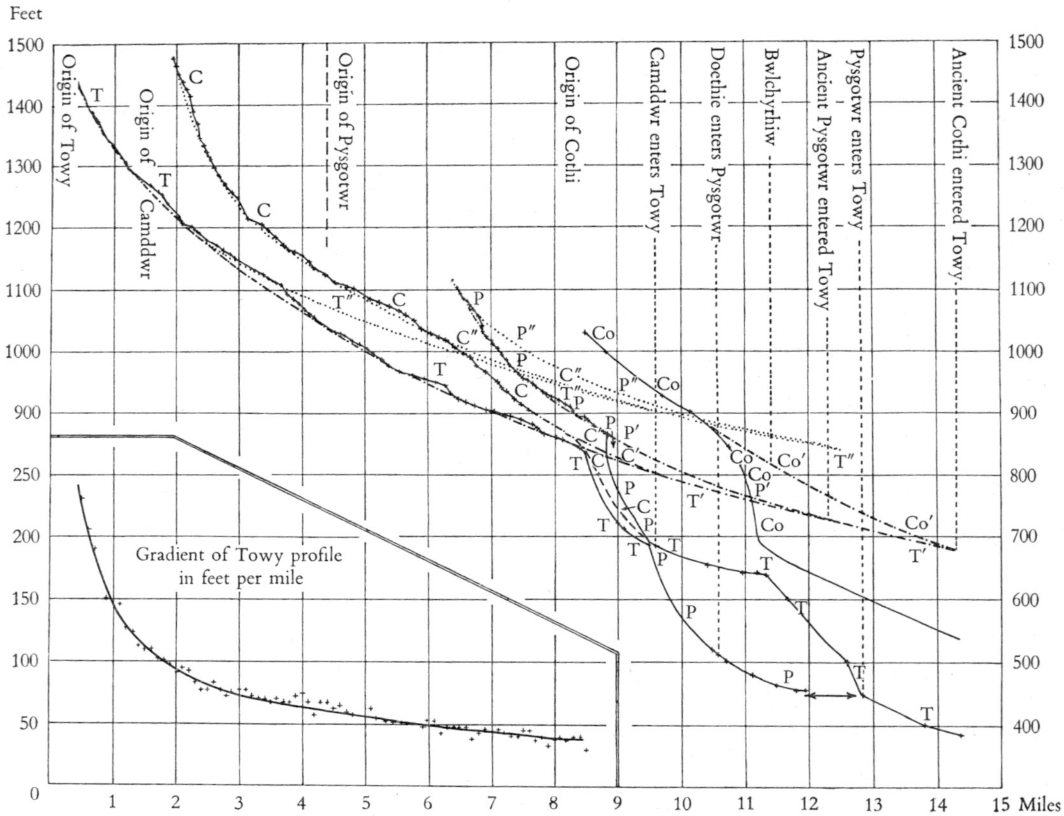

Figure 4. Profiles of the Towy Valley (from O. T. Jones, 1924). The points marked with a cross have been determined by levelling. The undashed letters refer to the present profiles, the single-dashed letters to the profiles at the Fanog base-level, and the double-dashed letters to the profiles

else I experience is consistent with glacial action on that rock, however indirectly we approach the matter.

But after Agassiz's[12] experiences in Switzerland and elsewhere, this mode of looking at objects became so ingrained that later he began to see evidence of glacial action in places where certainly there has been none, as in the Tijuca hills behind Rio de Janeiro. Such 'pictures in the fire' do not show consistency with our other mutually consistent impressions of the natural world. The importance of the self-consistency of our impressions about the natural world is fundamental in scientific investigation.

What I have tried to show in the foregoing geological instances is that we cannot be content with weeding out those parts of the science which are quantitative and which involve exact measurement and term the rest descriptive science. Very precise measurement plays an essential part in the study of the chemistry of igneous masses, and in radio-active estimates of the age of rocks, or in the extremely interesting study of gravitational anomalies, with the light they throw on the structure of the earth. But even when things of this sort are left out, the word 'descriptive' does not display the content of the rest of the science. The investigation of the Carmarthenshire plateaux was much more than mere description. Although it cannot be called a quantitative study in the generally accepted sense, it has a great deal in common with studies of that kind. In the case of the plateaux, what is really being investigated is the class to which a natural phenomenon belongs. That is essentially what was done by Lucas and Adrian in their investigation of the nature of the nervous impulse; equally it is what the physicist does when he investigates the nature of sound. That there are classes into which natural phenomena can be placed—and far fewer than our naïve expectations might lead us to suppose—is one of the essential characteristics of the natural world. Thus 'classification' has a much greater significance than the mere grouping of similar phenomena with the provision of appropriate labels. Later I shall try to show that this is also true of the taxonomic classification of animals and plants. Grouping and labelling are extremely useful when we are using 'classifiability' as a tool in some other investigation; for instance, it is essential to know the species of a disease-carrying insect such as a mosquito or tsetse fly when we attack the difficult ecological problem of its control; but in all the sciences, the determination of the class to which a phenomenon belongs is one of the most important conclusions of a piece of scientific research. In this sense all the natural sciences are classificatory.

Likewise, it has been said in the past that geology is an observational science in contrast with the experimental sciences. To quote from a small textbook of Jevons: 'To observe is merely to notice events and changes in the ordinary course of nature, without being able, or at least attempting to control those changes... In experiment on the contrary we vary at will the combination of things and circumstances and then observe the result.' In experiment we most commonly fix all the conditions we have succeeded in perceiving to be relevant, and then vary one. By this means we either relate the observed effect to the variation of the single condition, or we see if the effect verifies a hypothetical prediction. It has of course long been realised that no hard and fast line can be drawn between observation and experiment.

Jevons[13] says:

It is usual to say that the two modes of experience are Observation and Experiment. When we merely note and record the phenomena which occur around us in the ordinary course of nature we are said *to observe*. When we change the course of nature by the intervention of our will and muscular powers, and thus produce unusual combinations and conditions of phenomena we are said *to experiment*. Sir John Herschel has justly remarked that we might properly call these two modes of experience *passive and active observation*. In both cases we must certainly employ our senses to observe, and an experiment differs from a mere observation in the fact that we more or less influence the character of the events which we observe. Experiment is thus observation *plus* alterations of conditions.

But, in observation, at the least we choose the moment to observe or to have our attention drawn to the phenomenon. There are indeed 'natural experiments' such as the occurrence of eclipses. In the restricted phenomena of the astronomer and with his restriction of position this may be important. Instead of fixing conditions in the laboratory, we may wait for the appropriate time, or journey to the appropriate place, where we will find the necessary conditions to be fixed by natural events to satisfy the conditions for drawing an absolute conclusion. But if we had had to rely upon the concurrence of natural events alone for the advancement of physical science, indeed we could not have got far. But in astronomy, in geology and in biology observation of natural events at chosen times and places can sometimes provide information as wholly sufficient for a conclusion to be drawn as that which can be obtained by experiment.

The objects of study of geologists and biologists are exceedingly varied. It is of course just this richness of their material which makes progress so much slower in these sciences than with the limited objects of study in the physical sciences. But this richness does carry one advantage. The extent and variety of phenomena presented to the geologist very greatly increase the frequency of 'natural experiments' available to him. That is indeed why so much of the work of a geologist is done in the field. He is in fact actively making a selection out of the innumerable phenomena before him to test his hypothesis. In the case of the Towy river system, by walking to each tributary we are allowed to make an independent check of the base-line which streams are approaching. Such observation is much more than 'mere' description, for, as I shall try to show later, perception itself depends upon training, past experience and contemporary hypotheses. It is not that perception is modified by these but that they are built into our perceptive machinery in the same way as ability to detect certain configurations can be built into an electronic device. As Charles Darwin[14] said in his *Letters*: 'About thirty years ago there was much talk that geologists ought only to observe and not theorise; I well remember someone saying that at this rate a man might as well go into a gravel pit and count the pebbles and describe the colours.' It is difficult to see how neutral observation can in fact exist, from the moment that we place restrictions upon how it is done.

The richness of the phenomena available to the complex sciences can give the scientist an immense power to solve problems by the selection of observations of very different sorts. One half of Darwin's argument for the evolutionary origin of species depended on the powerful inductive argument he produced by the selection of an immense number of instances from geology, geographical distribution, morphology and embryology.

It is in fact the richness and complexity of their phenomena which distinguish sciences such as biology and geology from the physical sciences. *Physics and chemistry have been able to become exact and mature just because so much of the wealth of natural phenomena is excluded from their study*. There is no need for the physicist as such to go to biology for data, until in the last resort he has to take into account the fact that the observer is a living creature. I would call such sciences 'restricted'.

In contrast, biology and geology are 'unrestricted'. Men of science devoted to these fields must be prepared to follow the analysis of their problems into every other kind of science. If they wish to advance their subject they cannot possibly say, 'I will not burden my mind with chemistry, physics, or anything but my special interest'. If they do that—and some try to do so, even at school—they can make no major advance in their science and at best become no more than a sort of card in a catalogue with certain technical information written on it for others to use. We can in fact partially justify the rough division of the sciences by the Royal Society into 'A' and 'B' if we say that it corresponds to a division into the restricted and the unrestricted.

Though in general we may say that the physical sciences are restricted and biology is unrestricted, this does not correspond to an invariable distinction. Some parts of biology are restricted—as in certain parts of taxonomy; whilst, though in the physical sciences nuclear physics can fairly be called a restricted science, meteorology is unrestricted. The physical steady state of the air, and of the ocean, is necessarily determined by the presence of oxygen and carbon dioxide. But these are the result of photosynthesis and respiration; so that the physical problem leads us into biology.

This distinction between restricted and unrestricted sciences has a bearing on the evolution of exactitude and quantitative precision in the sciences. The more a science is restricted in the classes of its objects the more probable it is that we can frame far-reaching mathematical hypotheses about it which can be tested by precise measurement. Such hypotheses attempt to supply a very rigorous specification of the properties of objects and phenomena. The more exact and the more far-reaching the quantitative predictions, the very much greater is the possibility of failure and therefore of the probability of correctness when such numerous and detailed hypothetical predictions are congruent with observation.

But manifestly the difficulty of construction of such hypotheses is least when the number and classes of objects and phenomena are most restricted. It is much easier to construct and test hypotheses about the conduction of sound than about the process of learning.

On the other hand, restriction of the class of objects considered also restricts our ability to represent the whole gamut of natural phenomena. It is true that in the unrestricted sciences our quantitative models are more vulnerable; the chance of unknown factors influencing the phenomena is so high. On the other hand, confirmation of a hypothesis is most spectacular when it comes unexpectedly from a distant science. That is most evident in the complex unrestricted sciences. In these, our hypotheses do not have the assurance of a prolonged deductive safe-conduct, as in those concerned with, say, the interaction of elementary particles: but they may make up for this by showing congruence with an astonishingly wide range of natural phenomena. Darwin's theory of evolution by natural selection was of this kind, and its truth is quite as convincing as that of many accepted quantitative theories in the physical sciences. If we find the remains of a geologically recent fauna and flora and our hypothesis about its origin is congruent with its stratigraphical position, with the quantity of ^{14}C as measured by its radiation, and with conclusions on other grounds about former terrestrial climates and so on, the assurance of the truth of our hypothesis has a greater intensity than would be the case if the evidence in its favour all came from one restricted field of knowledge. Moreover, any systematic restriction in the scope of a science necessarily introduces a systematic bias in the picture of nature which can be derived from it.

There is another feature of the geological instances we have considered which invites discussion. It is particularly evident in the unrestricted sciences that the goals of scientific investigation may be at many different levels. When considering the Towy river system and the Carmarthenshire plateaux, it is to be noted that all the significant phenomena are on a large scale and, unlike the large phenomena of the astronomer, their interpretation does not seem to demand indefinite refinement of measurement. The conclusion that on the plateaux the gradients of valleys are converging to a level some 400 feet above the present sea level depended in the first place upon qualitative impressions of the eye. These were then supported by a survey of the river valleys. If the survey had been very inaccurate or made at very few points along the valley floor it would not have been of value: indeed the necessity for making the survey arose from the fact that contours established by the Ordnance Survey were too few. The surveys of Professor Jones were made at points which averaged about one tenth of a mile apart. A survey on this scale gave all the information that was wanted. It showed that the valleys were following a curve which could be represented by an empirical formula which by extrapolation in each case converged towards a base-line of about 400 feet above sea level. So far as the establishment of this characteristic of the valleys was concerned, nothing whatever would have been gained by surveying a thousand or a million or more points in each mile to an accuracy of a millionth of a millimetre in height. Not only are the plateaux and valleys themselves large things but our analytical description of them has a limit below which the information obtained does not help us with the problem. I do not mean by this that all information which may help us to reach a conclusion about the origin of the plateaux is necessarily to be got only from analyses to about a tenth of a mile in size. It might turn out that minute chemical analysis of the deposits could give information which would bear on the problem. But even here there would be a limit in size and accuracy beyond which no further useful information would be gained.

The same sort of thing could be said about many other systems. If we had never seen a petrol engine before and wished to determine how it worked, our investigation would be concerned with quite large objects like cylinders and sparking plugs enduring for a considerable period of time. Simply to determine the class of machine to which a petrol engine belongs, nothing would be gained by subjecting its parts to the most minute chemical analytical methods available to us.[15] It is course true that when we have understood how petrol engines work we can go on to other and more minute questions such as the optimal chemical composition of the fuel or the crystallographic properties of the metals of which the parts are composed. But even with these new problems we again achieve our goal at some level of size and duration.

The large extension in time and space of the phenomena with which a geologist is concerned and the gross level which suffices for analysis to proceed to answer the questions of his research seem to contrast strikingly with what is being done today in nuclear physics, with its continued analysis to events of ever smaller size and briefer duration. At first sight the concern of sciences like geology with gross phenomena might be taken to be simply an expression of immaturity in the state of their development. I do not think that is so either in geology or in biology, perhaps in any science. I will only say now that some most important things, such as the class of machinery to which the nervous organisation of our brain belongs, are to be settled by considering levels of organisation far above those which the nuclear physicist is now considering. Undoubtedly one of the essential features of central nervous machinery is the pattern of neurone connections; and though neurones are small to our eyes, there are levels of structure far below them, investigation of which may perhaps have nothing to contribute to answer the question, 'To what classes of phenomena do memory and learning belong?'.

In living tissues, however, there are special reasons why goals of investigation should often be just at a molecular or perhaps at an atomic level. Because our bodies are so much larger than the molecules of which they are composed it may appear to us that biological analysis is being pursued to indefinitely smaller and briefer things, made evident to us by the scalpel, the light microscope and the electron microscope; and by clocks, camera shutters and electronic devices. But I think this is due to the fact that the really important structural features which distinguish living from non-living matter happen to concern events which spatially are just about the size we can see with an electron microscope and which extend in time from enormous periods of years down to rather less than a millisecond—just as the events which occupy the attention of a geologist are often as large and as enduring as one of the Carmarthenshire plateaux.

If we consider the whole range of the phenomena of the natural world, we can decide to study events which are parts of it, or which are parts of these parts, and so on indefinitely. If we choose our divisions appropriately we find at various levels of division events and objects associated with such events which are worthy of study and which do not necessarily require analysis to lower levels by unlimited further division. Indeed, what is important can easily be lost in such an analysis, just as the significance of a performance of a play like *Macbeth* would be lost if all we saw was a part of a part of one scene.

Whitehead,[16] when discussing the ideal concepts of nature at an instant with no temporal duration and points with no spatial extension, notes how fundamental such concepts are for the expression of physical science. When we consider the particular events of the natural world, that is, the specific character of a 'place' through a period of time, we can approach the limiting ideals of point and instant; but they remain simply as ideal limits to which we can converge by a process of successive abstraction. He says:

For example, we see a train approaching during a minute. The event which is the life of nature within that train during the minute is of great complexity and the expression of its relations and of the ingredients of its character baffles us. If we take one second of that minute, the more limited event which is thus obtained is simpler in respect of its ingredients, and shorter and shorter times such as a tenth of that second, or a hundredth, or a thousandth—so long as we have a definite rule giving a definite succession of diminishing events—give events whose characters converge to the ideal simplicity of the character of the train at a definite instant. Furthermore there are different types of such convergence to simplicity. For example, we can converge as above to the limiting character expressing nature at an instant within the whole volume of the train at that instant, or to nature at an instant within some portion of that volume—for example within the boiler of the engine—or to nature at an instant at some point of the train. In the last case the simple limiting characters arrived at will be expressed as densities, specific gravities and types of material. Furthermore we need not necessarily converge to an abstraction which involves nature at an instant. We may converge to the physical ingredients of a certain point on the track throughout the whole minute. Accordingly there are different types of extrinsic character of convergence which lead to the approximation to different types of intrinsic characters as limits.

It seems to me that this throws very valuable light on the way we construct our mathematical models of the natural world. But there are other interesting features of such 'abstractive sets'. The event which is the life of nature of the Carmarthenshire plateaux during one minute or during many years is certainly complex and much of it baffles us; but not all. We can perceive enough of certain of its relations and ingredients to perceive that there is a scientific problem to be attacked and that the problem will not be furthered by continuing the process of abstraction indefinitely. When we are told, 'An elephant weighing four tons slides down a grassy slope at an angle of 45°: the coefficient of friction is so-and-so', it is usual to dismiss the elephant and grass. But that is just what we must not always do.

When we consider events in this way we must not always treat them as though they were of some uniform substance which can be indefinitely divided with impunity. Properly chosen an abstractive set derived from an event may pass through several stage of highly individual character which are worthy of study witho further abstraction.[17]

SUMMARY

1. In our everyday consideration of them, the sciences are apt to be taken as though they could be placed in a linear series. Mathematics and physics are at the top and the others are arranged down the rungs of the ladder up which they are proceeding as they become more exact—as in a dice-game. So-called descriptive sciences, such as taxonomy, stand at the bottom, still waiting for a lucky throw.

2. In fact the relation of the sciences is that of a multidimensional network with cross relations, as illustrated by geology and biophysics. Only an arbitrary division of the sciences is possible, though for practical reasons one must be made. The division into physical and biological sciences is such a division of practical convenience and no more.

3. Nevertheless, one can see a general decrease in complexity as one passes from geology or biology to physics, though the complexity is not of the same kind in all the complex sciences. The simplicity of the physical sciences, combined with the attention given to them because of their immediate powerful technical consequences, is the cause of their rapid success. Intellectually magnificent though the attack on them has been, the problems they present are easier than those of the geological and biological sciences.

4. There is one real, and graded, distinction between sciences like the biologies and the physical sciences. The former are *unrestricted* sciences and their investigator must be prepared to follow their problems into any other science whatsoever. The physical sciences, as they are understood, are restricted in the field of phenomena to which they are devoted. They do not require the investigator to traverse all other sciences. But while this restriction is the basis of their success, because of the introduction of this restricted simplicity of their field we cannot necessarily take them or their methods as typical of all the sciences.

5. The fundamental contrast is not between biological and physical sciences, but between *unrestricted* and *restricted* sciences.

6. Geology deals with complex physical systems and a described example shows the similarities to and differences from what we ordinarily understand by physics.

7. The chosen example shows that neither 'descriptive/exact' nor 'observational/experimental' is an exact definition of the difference between geology and physics.

8. Complex sciences, because of their complexity, offer so many 'natural experiments' that the disadvantage of not being able to set up controlled experiments in a laboratory may be largely offset.

9. The example chosen, of the origin of a river system, shows that the scientific objective, or goal, concerns what is large in size and duration. The advance of contemporary physics seems at first sight to suggest that the goals of all the sciences are reached by indefinite analysis to smaller and briefer phenomena. This does not in fact cover the goals of the field geologist. His conclusions are complete in themselves, and more minute analysis does not affect his conclusion though it may raise other problems.

10. But these other problems also have analytical goals which may be at a smaller and briefer level of phenomena, beyond which nothing is to be gained about the immediate conclusion by further analysis—though again that may raise fresh problems.

11. The position of the biological sciences is deceptive because the goals of investigation so often concern molecular events.

2

The features of the natural world

IN THE LAST CHAPTER we discussed different kinds of natural phenomena of different levels of size and duration. We examined some of their features and how these imposed differences on the character of the various sciences and on the goals of their scientific investigation. In all such investigations, whether in the field or in the laboratory, the scientist takes it for granted that his perceptions give him direct information about a real world external to himself and which includes his own body. Indeed he does more than take it for granted: for that might imply some continual intellectual acknowledgement of what he is doing—whereas in fact he just behaves that way. He does this just as surely as the ordinary man in the street or as a philosopher when he is catching a train. I am far from competent to discuss the old and exceedingly difficult question of the relation of what we perceive to supposed reality. But there is one aspect of it that I must discuss: the mutual consistency of all our perceptions about the world that we call real. On this all our scientific observation and interpretation is based. I do not mean by this that some of the models by which we describe phenomena do not involve paradoxes, such as the wave–particle dualism, by which we have to describe an electron sometimes as a wave and sometimes as a particle, two models which would be inconsistent at our gross level of perception. I do mean that if I say, 'There is a tomato', unless I have made an error, or am undergoing a hallucination, all my other impressions are and will be consistent with that assertion, however far we pursue them. It is characteristic of such an assertion about the real world that it is consistent with all our varied impressions, from touch and from the indescribable proprioceptive and labyrinthine impressions about spatial position and orientation. It is consistent with our perceiving a column of ants attacking the tomato and carrying pieces of it away; or with a small boy throwing it at a wall and smashing it; or someone saying, 'Let's have a salad, anyway we have a tomato'. If I say I 'truly' perceive that I have a tomato in my hand I cannot pursue all my other impressions and arrive at the conclusion that it is not there or that it is a sixpence. Hallucinations[1] are distinguished by us from reality by lack of this mutual consistency. They may for a while seem to show consistency, but sooner or later they will come into collision with the rest of our experience. I acknowledge this in common speech. If I suspected that what I saw was a hallucination I might say, 'I pinched myself to see if I was awake'—that is, I tried to test the consistency of these perceptions with others.

In everyday life we are indeed often in some error in our perceptions. Usually we must stay content with inadequate information. Events force us to act, and will not stay for an answer. If a glance indicates that tomatoes are on the kitchen table and this is consistent with the fact that someone said he put them there, we accept this; that is, we give unconditional assent to this—even though our assent may later be shown to be falsely based because someone played us a trick with wax fruit.

Fortunately our perception seems instantly to apprehend many of the relations of things in the real world with a high probability of absolute correctness; the more of these relations that are consistent with the object of our attention, the very much greater is the probability that our perception is not in error.[2]

Our perceptions of objects depend in a high degree upon the learning of their relationships. We can perhaps think of objects as envelopes of perceived relations, using the word 'envelope' rather in the same way as a mathematician speaks of a figure being an envelope of a set of functions. Very remarkable results have been obtained from the study of persons who, by reason of congenital cataract or other defects of the lens and cornea of the eye, subsequently cured, are given the sense of sight late in life.[3] At first there is a meaningless welter of lights and patches. Learning to perceive these as figures and objects and to relate these to touch and proprioception may be fairly rapid or very slow. It takes time. But once learned, we no longer live in a world of randomly related lights and patches, but in a world of perceived objects. We can only get back to the lights and patches, if at all, by a conscious intellectual process. Our perceptive machinery, whatever its nature, has been changed by our experience. This same refinement of perception increases in our normal life when we learn about objects. If I say, 'There is a shore-crab, *Carcinus maenas*', what I perceive is far richer than was the case when I was a child and had not studied crabs. I do not see the crab as I did as a child and then argue about it with my acquired knowledge—I just see it at once as a naturalist who knows species of crabs. The same is true of a collector of porcelain who sees a piece of Worcester of Dr John Wall's period.

Past experience has 'built in' to the receptive system all manner of special devices—and in a sense these are 'learned'. But in operation with this 'hotted-up' machinery of perception the 'whole' object is perceived at once and in relation to the whole sensory field, and the machinery and the conditions of existence are such that this is possible.

The tacit suggestion that in all our varied perceptions there is one and only one set or aspect of them in which they are mutually consistent is the essential basis of our acceptance of the real world. Among its consistencies are the perceptions that other people, and also animals, react to objects as we do; both even show similarities of aesthetic preference. I do not see how this assumed consistency of our impressions can be justified. At any moment, for all I know, the world may become crazy and reliance on the former consistency would be misleading. But among my impressions of the world it seems to me that natural selection takes a heavy toll when an organism appears to make a mistake about what in everyday speech I would call a real object in the natural world.

The reason that I have made this digression about perception is because our scientific attack rests upon the acceptance of a real world, to be made evident by the self-consistency of our set of perceptions. In doing this we judge a perception against a background of all our other perceptions and what we have learned in the past about them. We have not had every conceivable kind of perception, many of the things we learn about their relations we learn indirectly on authority, and the authority may be in error. When this happens, there may come a time when some new conclusion—such as the mutability of species through evolution—drastically alters the whole system of the images in our minds by which we have come to represent the natural world and against which we judge the consistency of our present perceptions.

As no more than an amateur philosopher I have been rather puzzled by the fact that when attempts are made to analyse perception and its relation to the supposed existence of a world of real objects, so little is said about this consistency of what we assent to as perceptions of real objects, and how much is said about sense-data: colour patches, pressures, a noise, a smell. My difficulty about sense-data is that the instances people give of them are at once so complicated and so inadequate. The physiologist warns us how complex is the stimulus of a tactile field. It is already a matter of relationship perceived. Even the discharge of a single touch receptor is not a simple measure but, through adaptation, becomes a partial differential of the stimulus. Both the psychologist and the physiologist warn us how much of what we perceive depends upon information about relations inaccessible to consciousness. How much do the indescribable proprioceptive and labyrinthine 'stimuli' contribute to our instant perception of an object and its orientation?

If our conviction of reality rests upon the mutual consistency of relationships directly perceived, perhaps there is but limited value in analysing perception in terms of sense-data. When considering perception, Dr Price[4] says:

When I see a tomato there is much that I can doubt. I can doubt whether it is a tomato I am seeing, and not a cleverly painted piece of wax. I can doubt whether there is any material thing there at all. Perhaps what I took for a tomato was really a reflection; perhaps I am even the victim of some hallucination. One thing I cannot doubt: that there exists a red patch of a round and somewhat bulgy shape, standing out from a background of other colour-particles and having a certain visual depth, and that this whole field of colour is directly present to my consciousness...And when I say that it is directly present to my consciousness, I mean that my consciousness of it is not reached by inference or by any other intellectual process (such as abstraction or intuitive induction) nor by any passage from sign to significate. There obviously must be some sort or sorts of presence to consciousness which can be called 'direct' in this sense, else we should have an infinite regress.[5]

Now it does seem obvious that something is presented to consciousness; but it is not clear that by transferring consideration from a tomato to a bulgy red patch we succeed in analysing the matter. Everything said in this passage about the tomato can also be said about a bulgy red patch. Its bulginess may be an illusion. It is not even a simple elemental quality but something arising from sensory relationships. The same is even true of the red patch. From the time of Helmholtz to the present experiments of Dr Land[6] on the illusory variety of colours that can be produced by achromatic illumination under different conditions, we have been warned that when we see a red patch there has already been some 'passage from sign to significate'. In an important sense the red patch is directly presented to our consciousness—but so also is the image we call the tomato. Both arise from relations which involve to some extent the whole sensory field, some of which may not even be accessible to consciousness. Our instant image of a tomato seems direct and not mentally built up by inference from sense. It requires, on the other hand, an intellectual operation to abstract sense-data from that image; and when we have done so we have still failed to disentangle elementary qualities. We have set out in the wrong direction.[7]

Is there perhaps danger if we look upon 'sense-data' as something from which we build up perception? Rather, perception of an object is the unit—if that is indeed an appropriate term—and sense-data are highly intellectual abstractions from it, robbed of those relationships which are the basis of our conviction of reality. Is not the situation just a little like that of Tommy Smith when he is sent to a party? He is scrubbed, washed and brushed and loaded with maternal admonition. Too many things has he got to, too many things has he not to. He has not enough degrees of freedom to display the young old Adam. Even so he may show awkward relationships with other objects—he may get jam on his front and pull Miranda's hair; but his parents must take the matter philosophically and say, 'Tommy is not behaving like himself.'

The scientist, then, is investigating what he believes to be a real world consisting of objects whose relationships he can study. By this, and by discussing them with other people and reading about them, he gets a contemporary description of this world. What is its nature as we see it every day? I mean the world of objects we see in the laboratory or outside; I do not mean to discuss, now, space, time and ultimate particles, but just to consider what are the objects we perceive in the natural world and how we interpret them.

The most striking feature of the everyday world of objects is the enduring character of the things in it. Apart from cyclical changes of day and night and the seasons, the world is full of things like tables, hills, valleys and the sea, which are there to be seen every day. I shall call these enduring objects. Some, such as frost on window-panes, are certainly ephemeral by our ordinary ideas of time. Others, like the 'eternal hills', far outlast the span of human life. Investigation shows that all objects of this kind suffer decay, sometimes gradually, sometimes abruptly. By a sort of natural selection our environment is a residue of the not-yet-destroyed.

Of course the frequency of occurrence of any kind of object depends not only upon its average endurance but also upon its rate of reproduction. Puddles of water are ephemeral but are not uncommon. But the things that strike us are the individual things that endure.

Particularly it is solid things that endure in this way. Gradually they suffer decay or denudation. That may be a gentle disappearance of ordered organisation, as in a bicycle left out in the open that, given enough time, will gradually become a pile of rust. Or rather striking secondary order may appear. The denudation of a land surface engenders river valleys, calcium carbonate may deposit as stalactites, mud as it dries may yield remarkable patterns of suncracks, agates may develop Liesegang rings. Nor must we forget crystalline structures and the varied forms they may give rise to according to their condition of deposition—from ice to quartz.

Some of the complex structures that arise in denudation are merely long-lasting. Others have a stability because small changes tend to bring the system back towards its former state. When a freshet clears sand from a region of the bed of a stream, it also brings down more sand to take its place. But here we are dealing with a different class of object.

Among the things that endure are not only long-lived structures, like hills, but also things like rivers, waterfalls, glaciers. In an important sense these are not permanent objects. They are dynamic equilibria in which their enduring form and properties are continuously maintained by fresh material. The grandest of these is our atmosphere. The air we breathe is as constant and necessary a component of our world as the ground on which we walk.

We owe our oxygen to its continuous production by plants. The carbon dioxide, apart from a small amount of volcanic origin, arises chiefly from the respiration of organic matter by bacteria, plants during darkness, and to a less extent by animals. Geologically speaking, the rate of turnover is rapid. There is even a detectable difference in the composition of the atmosphere during the longer light and darkness of polar summer and winter. Calculation suggests that if all production of oxygen were to cease and its consumption to continue unchanged, the supply in the atmosphere would only last some thousands of years. Nevertheless, an oxygenated atmosphere must have been maintained on this earth for at least 500 million years and the appearance of ferric rocks at a certain point in the succession of pre-Cambrian rocks suggests that it has been maintained for far longer than that.

Thus the atmosphere, which is but a dynamic equilibrium, is a most enduring feature of our world. The combination of enduring objects and enduring dynamic equilibria in the world around us gives us a remarkably stable natural world against which to judge the phenomena which we perceive and to which we can refer our perceptions for their consistency with reality. It is difficult to see how natural science could have advanced at all if its phenomena took place in a continually changing and unstable world.

In our perception of objects it is important to notice that our senses only cover a certain range; our immediate perceptions are limited by what I may call our 'sensory spectrum'. We only see with light over a certain range of wavelengths. We cannot see radio waves. Curiously enough, a great many animals behave as though they were responding to a sensory spectrum very like ours. Some butterflies, like the cabbage white, behave as though they saw the same colours in the same spectrum with the same limits as we see them, and, moreover, experiments suggest that their colour vision is liable to the same errors of colour contrast that we find in our own eyes. In contrast, however, some animals have significant differences in their sensory spectrum. The pits in the head of pit-vipers are a sort of pinhole camera by which they perceive infra-red radiation from the warm bodies of small mammals on which they prey. Bees are notorious in that their 'colour sense' is both well developed and corresponds with quite different divisions of the visible spectrum from our own and, moreover, it includes a region of extreme sensitivity in the ultra-violet, apparently used particularly in connection with orientation of their flight in relation to the sun and sky.[8]

But it is not only the restricted sources of information which are important in perception, but also the magnitude of the characteristics of the objects we see. With our unaided senses we do not perceive things which are too small or too vast or which endure for too short a time. The endurance of an enduring object is to be measured against the time-scale of our own lives, and so on. Events which are too small, too large, too quick, or too slow are not perceived, and unless our attention is drawn to them by indirect means we know nothing about them. For our purpose the importance of this is the obvious fact that one cannot make a scientific hypothesis about something one is not aware of.

We become aware of phenomena outside the range of our sensory spectrum by their indirect effects within it. And we do this particularly by means of instruments which bring dimensions into our range of appreciation. Such are microscopes, telescopes, or oscilloscopes, which render visible changes that take place at very high speed. Or there are ingenious devices by which speech or the song of birds can be reproduced as visible patterns, thereby enabling us to perceive characteristics of it not available through our auditory sense; or we can record the slow oscillations of earthquakes or of ocean waves and, playing these through at a high speed, we can appreciate what has happened by detecting audible patterns of tone or visible oscillations of particular frequency.

The natural world as we see it every day shows us matter in three states. We are not generally concerned with that new fourth state, the 'plasma' of physicists. Most complex objects and equilibria involve solids.

Experience indicates that, left to themselves under the gravitational field of the earth, the material components separate out with the solids below, the liquids such as water in a position of least potential energy in their hollows, and the gaseous atmosphere above. This state is disturbed by the continual receipt of energy from the sun. The general rules which govern the distribution of energy in material systems were stated by the physicists of the last century. They gave us the first two laws of thermodynamics: the principle of the conservation of energy and the principle of the degradation of energy. This second law of thermodynamics describes the fact that in any complete physical system the total amount of disorder tends to increase; or that a quantity, entropy, which is a measure of disorder, tends to increase. You cannot tap the heat energy of a closed system already in equilibrium and convert it into work, for that would mean decreasing its disorder by requiring the system to separate itself into two temperatures; a phenomenon of indefinitely high improbability. There is a third law, that of Nernst, which says that at the absolute zero of temperature the entropy of any system becomes zero. I am not at present concerned with this law. The important one which describes the behaviour of most of the ordinary physical world as we see it is the second law of thermodynamics. It is this that gives direction to all that goes on. A film run backwards, with time apparently reversed, looks absurd, strange and magical and curiously prophetic in just those features which illustrate the second law. Where, as in the revolution of the planets round the sun, we are unaware within the span of our lives of the degradation through radiation and tidal and other friction, a film of the phenomena is perfectly reversible and the reversal of time causes no comment.

The second law tells us of the direction in which change is taking place: in a complete physical system it is always running downhill into more disorder. All kinds of energy, such as excess motion in one part of the system, are ultimately distributed through the whole of it, just as a hot body loses heat until it and its surroundings have reached the same temperature. Now at first sight this sort of change seems to contrast strikingly with what we see in the most interesting of all the objects we perceive, living organisms: and it is an attempt to see the difference between living and non-living systems which is ultimately going to be the object of my discussion.

Living organisms are obviously ordered structures and, what is more, their whole existence seems aimed at producing more of these ordered structures. True, the beautiful order of a living thing vanishes away, but new generations continually arise and grow to recall its prime. Moreover, the general impression we get of their evolutionary history is that, as time goes on, their degree of order has in some not easily definable way increased till it corresponds to the difference in degree of ordered structure and function between a man and what may originally have been like a very simple bacterium.

Notwithstanding all this, no contradiction has ever been established between the general laws of physics and chemistry and those which hold for living systems. From the beginning of the last century it became increasingly apparent that 'organic material' did not owe its peculiarities to vital principles peculiar to it; and perhaps the most striking feature of the physiological analysis of the activities of living creatures was the continual proof that if we consider the whole system of the combustion of the food they use as fuel and the mechanical and other work done, as well as that for all else they are doing, the thermodynamical rules are absolutely obeyed, as far as accurate quantitative measurements can tell us. The one thing we can say is that the increasing organisation of living structure is built up at the expense of still more increasing disorder elsewhere; so that though we have no evidence in the whole system that the second law is contravened, in a sense living organisms grow by accumulating order or negative entropy; and like certain machines they do this by guiding the increase of a much larger quantity of entropy. Like the machines, they are parasitic on the increase of entropy, but, unlike them, the result of their activities is to produce more similar machines. Organisms delay and guide the degradation of energy to give limited amounts of ordered system which do this very thing.

The second law tells us nothing about how quickly or how slowly energy will be degraded, or by what routes it will travel, or what will be the consequence of its travelling by these routes. These matters, however, are clearly important in the description of what characterises a living organism. But the reason I raise the point now is because I am not convinced that these essential characters of living organisms are peculiar to living organisms. I can make my difficulty clearer if I take the following comment of an eminent physicist about the contrast between living and non-living systems.

Professor Schrödinger in his little book *What is Life?*[9] deals with the matter as follows:

What is the characteristic feature of life? When is a piece of matter said to be alive? When it goes on 'doing something', moving, exchanging material with its environment, and so forth, and that for a much longer period than we would expect an inanimate piece of matter to 'keep going' under similar circumstances. When a system that is not alive is isolated or placed in a uniform environment, all motion usually comes to a standstill very soon as a result of various kinds of friction; differences of electric or chemical potential are equalised, substances which tend to form a chemical compound do so, temperature becomes uniform by heat conduction. After that the whole system fades away into a dead, inert lump of matter. A permanent state is reached, in which no observable events occur. The physicist calls this the state of thermodynamical equilibrium, or of 'maximum entropy'.

And then again:

It is by avoiding the rapid decay into the inert state of 'equilibrium' that an organism appears so enigmatic; so much so, that from the earliest times of human thought some special non-physical or supernatural force (*vis viva*, entelechy) was claimed to be operative in the organism, and in some quarters is still claimed.

My difficulty is that almost everything that is said here about life could at least in some measure be said about a thunderstorm. That goes on 'doing something', moving, exchanging material with the environment, and so forth, and that for a much longer period than we would expect an inanimate piece of matter to 'keep going' under similar circumstances. It is by avoiding the rapid decay into an inert state of 'equilibrium' that a thunderstorm appears so enigmatic: so much so that in earlier times our ancestors saw in it Thor with his hammer on the anvil. And if we may consider gentler and far more enduring systems, much the same things can be said about the Carmarthenshire rivers.

The important phrase is '...than we would expect in an inanimate piece of matter'. I suggest that the mental image in this case was something like a laboratory in which there is a piece of metal at a temperature T_2 which is following Newton's law of cooling towards the temperature T_1 of the surrounding room. For our mental and experimental convenience we choose simple systems of this sort for our investigations. Mathematical relationships which describe what happens are then more easily within our grasp. The laws of thermodynamics were themselves perceived by considering such homogeneous systems. No physicist in his senses would begin exact physical investigations about the nature of matter by a study of the Towy river system. The laws which he discovers are true and hold generally. But we must bear in mind that a very important concession has been made to the frailty of our understanding by choosing out of all the rich variety of nature what seem to us to be the very simplest phenomena for our first investigation. No harm is done by this inevitable method of our scientific procedure provided we bear in mind that we have ourselves put this element of simplicity into our view of nature and therefore we must not be too surprised to see simplicity in the answer. As soon as we go out of our laboratory and put up our umbrella we are dealing with a more complicated world even in physics.

If to a complex physical system of solid, liquid and gaseous objects energy in some form is applied, there may be several distinct pathways along which it will travel. This really follows from the important physical principle of Le Chatelier,[10] 'the principle of least action'. If to a system in equilibrium a constraint be applied, a change takes place within the system tending to nullify the effect of the constraint and to restore the equilibrium. If I have a closed system of water vapour over water and I apply some heat to it, some of the heat energy will be taken up by the conversion of some more water into water vapour. But if now we allow heat to escape, it will not only be supplied by the simple fall in temperature of the system, but some will be lost by re-condensation of water with the liberation of stored heat. Such a secondary route for the degradation may be a very slow and much delayed process and can take the stored heat energy through un-expected channels. Most of the sources of energy which we tap for industrial purposes, wind, water-power, coal and petroleum, perhaps even nuclear transmutation, are ultimately derived by different paths of degradation on the route of the conversion of solar radiation into heat.

Depending on the physical conformation of the system, the energy can travel through a variety of paths with remarkable consequences. If the water in a reservoir at a certain height is allowed to flow to a lower level, the gravitational potential energy is converted into heat. In this final condition the chance that the random motion of the water molecules will so conspire as to make a single drop leap upwards to a higher level is utterly remote. But by means of a suitable physical structure, a hydraulic ram, the water flowing down a pipe from the upper reservoir is allowed over a period of time to convert potential into the kinetic energy of a moving column of water. A suitable automatic valve, by suddenly stopping this flow, can now enforce the conversion of kinetic energy into high potential energy of pressure in a small volume of water, so that this small quantity can even be raised well above the level of the upper reservoir.

In the case of the hydraulic ram we have a physical system which can increase greatly the order of a small part of the system at the expense of still greater disorder in the remainder. Moreover, we can very considerably delay the conversion of potential energy of this part of the system into heat. The ram differs from a living organism in that the degradation of this energy is not used for the maintenance of the physical structure of the machine. But there are natural systems which show even this property: again, thunderstorms are just such an example. The parallels between living organisms and thunderstorms and a variety of other meteorological phenomena are remarkable. They differ in that thunderstorms arise only by spontaneous generation. Since they are incapable of sexual reproduction, natural selection can only act by selecting individuals and not by acting upon the whole species. Like living organisms, they require matter and energy for their maintenance. This is supplied by the situation of a cold air-stream overlying warm moist air. The situation is unstable and at a number of places vertical up-currents occur. These, once they develop, become self-maintained through the liberation of heat consequent upon the formation of rain as the warm damp air rises. Each up-current 'feeds' upon the warm damp air in its neighbourhood and is thus in competition with and can suppress its neighbours—a familiar ecological situation among animals. The storm is parasitic on the increase of entropy which would result from the mixing of warm moist and cold air to form a uniform mass.

The storm itself has a well-defined morphology of what can

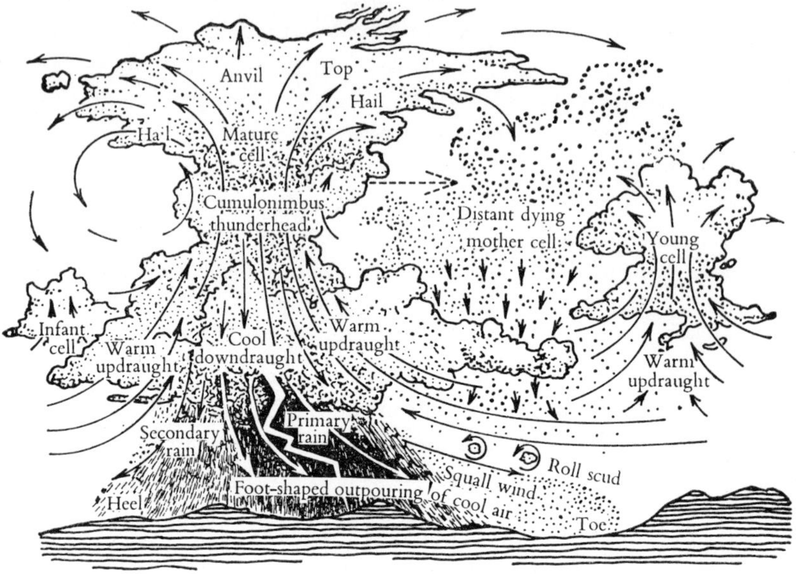

Figure 5. Diagram of the organisation of a thunderstorm as a system of individual cells. As the figure shows, the storm has 'a well-defined morphology of what can almost be called functional parts'. As the descending central downpour of rain increases, the mature stage of the original cell passes into the dissipating stage. This presently culminates in the total cessation of up draughts. The process is meanwhile being repeated in the offspring thunderstorm cells and cloud towers, each of which in turn passes through the same process of 'development, maturity and decay'. (From G. Murchie, 1955, *The Song of the Sky*, p. 202. London: Secker and Warburg; Boston: Houghton Mifflin Company.)

almost be called functional parts. An isolated description of the vertical up current would allow one to narrow down the class of system of which we could predict it must be a part, and that class would include thunderstorms. This comes fairly close to a characteristic of the functional anatomy of organisms that from a given part we can infer its function and make predictions about its relations to the rest of the organism.

The electrical phenomena of a thunderstorm are remarkable as an illustration of the fact that in a morphologically complex

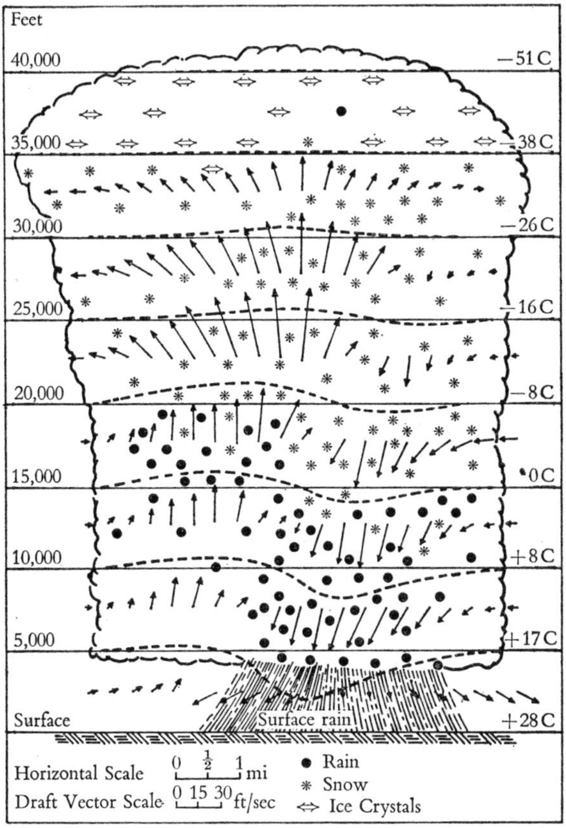

Figure 6. Sketch of a vertical cross-section through a thunderstorm cell in a mature stage, showing vectorially the air circulation. The temperature distribution shown is typical of summer thunderstorms in the eastern United States. (From H. Byers, 1949, *Science*, **110**, 293.)

system even a small difference in the temperature of two masses of gas can by indirect paths of energy give a small mass of gas at the exceedingly high temperature of a lightning flash. But these electrical phenomena of the storm are not of functional importance to it. They resemble the concomitant properties of physiological systems—like the colour of the hair in mammals. But in thunderstorms, in the absence of sexual reproduction and selected inheritance, no functional value can become attached to these concomitant properties—as contemporary literature would lead us to believe it is to blonde hair.

'Organismal' characteristics can be noted in rivers as well as in meteorological phenomena. In fact, while it is indeed a characteristic of living organisms that in a sense they evade the decay to thermodynamic equilibrium, essential features of this condition are to be seen in some purely physical systems of our everyday experience. It may be asked how it is that in this world of ours systems of this sort come to exist. I will only point out that provided such systems are possible, they are likely to become features of this world simply because, by delaying the approach to equilibrium, they necessarily endure. The important fact is that the variety of nature is such that they are possible.

Although living organisms are not the only things in this world that are characterised by the fact that they are 'dynamic equilibria' through which matter and energy pass, and in which energy, directed during its degradation, results in some local increase in order within the system, yet it is quite true that no other things show this property in so high a degree as do living organisms. It is indeed to be seen in a thunderstorm, but it is much more strikingly evident in an amoeba. I want now to consider what other special characters strike us about living organisms to see if we can decide if they are unique.

Professor Woodger[11] recently called attention to a basic fact about living things in that they have parts which stand in a relation of existential dependence to one another. That is, there are clearly defined living parts, such as the head or limbs of a man, which depend for their continued existence upon the remainder of the body of which they form a part. A second feature of importance in living organisms is their hierarchical character. Cells divide to give cells, and so on. In the cellular organisms the cells undergo differentiation to give a complete organism. The same kind of process takes place at many levels. By various methods organisms reproduce more organisms. Similarly groups of organisms can form colonies with separate or connected individuals. By various

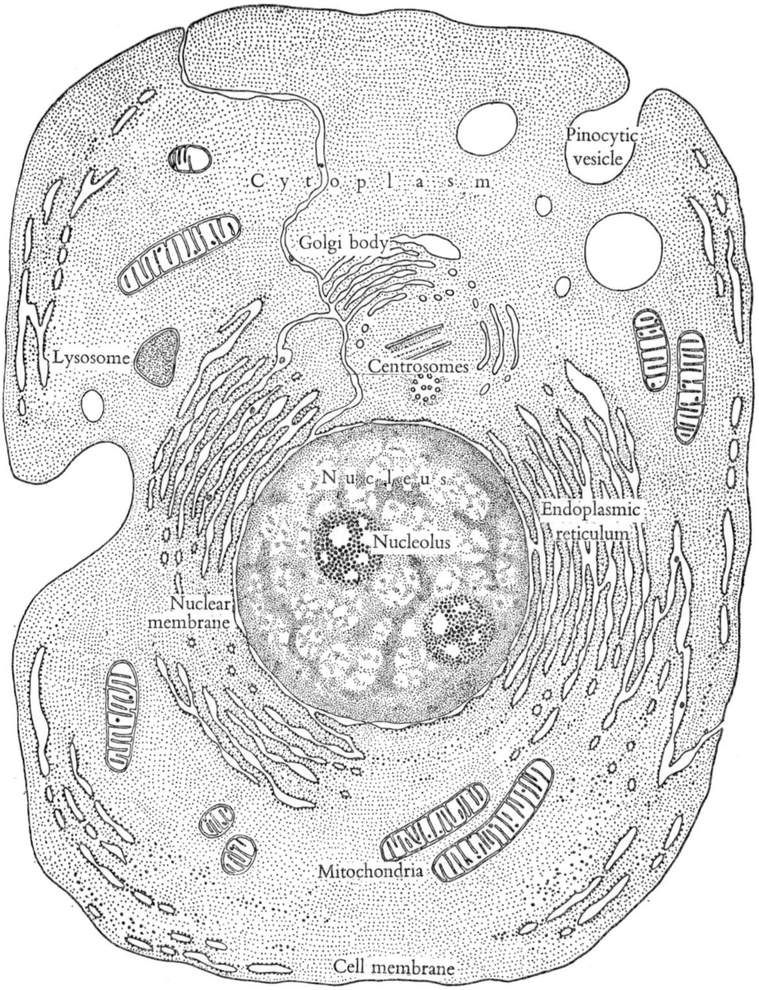

Figure 7. Diagram of a cell showing the organelles which form essential parts of its machinery. (From J. Brachet, *Scient. Am.* September 1961, p. 55.)

methods these colonies can also reproduce. Going in the opposite direction, various 'organelles' of the cells, such as mitochondria, plastids, and particularly the nucleoprotein structures, the chromosomes, undergo reduplication. All living systems, cells, organisms, colonies, can reduplicate themselves and reduplication can proceed without limit: it does not gradually run down. Such completely self-perpetuating hierarchies of generations from an original cell, organism or colony have no precise parallel in the inorganic world. For all their organismal characters, thunderstorms do not reproduce in this way. It is true that in a natural transmutation series of radio-active elements we have a hierarchical division of atoms into other structures which may then divide again, and so on. But the processes of division in each generation are not wholly equivalent to each other and cannot continue indefinitely.

Some other complex physical systems show a more complete resemblance to living ones. A drop of ink allowed to fall into a glass of still water sets up a vortex ring. As the ring descends it expands and generates a number of subsidiary rings. These descend and expand like the parent ring, and in due course they may produce a third generation of vortex rings, and so on. There is an obvious parallel here with the successive generation of living cells by division. But such a vortex ring system lacks an essential feature. There is no continuous production of fresh material of a like kind, so that the rings get smaller and smaller. Likewise there is no addition of energy to the system, so that finally through friction the rings cease to exist. If self-developing systems like thunderstorms could undergo successive reproduction in this way, the parallel would be very much closer.

The possession of existentially dependent parts is an exceedingly important character of living systems.[12] On it is based the functional character of their parts. With this goes teleological description; and though final causes necessarily come in for a good deal of criticism, this way of describing structures is not without its importance in biological research. Characteristically, of a living organism we can ask, 'What is this part for?', and we expect a useful answer. We cannot usefully ask that about the parts of an atom or a molecule—unless we are prepared to see design in inanimate structure, a point I do not wish to discuss now. We can, of course, ask what a part of a 'machine' is 'for'. If we find a carburettor on a scrap-heap, we can usefully ask what it is for. From such a part in isolation much can be predicted about the machine of which it is a part. Indeed detection of function of this sort was not unimportant in some operational research during the war.

To a limited extent we can expect a useful answer if we ask what the vertical current which generates a thunderstorm is 'for'. But the question is far less useful than in the case of machines and living organisms. Spectacular predictions of this sort have been made even from a part of a single bone.

In living organisms we can go further than this. Their parts can be repeatedly subdivided into dependent parts which retain the characteristic of being functional: fore-limb, hand, digit, cell. Unlike machines, or at least unlike those we have today, in living organisms we can pursue this functional division almost down to a molecular level. Clearly there must be a limit to this when we subdivide the part down to the molecules or even to the atoms of which it is composed.

An illustration of these principles can be seen in some parasitic organisms. Certain parasitic Crustacea (*Sacculina*) have a larval stage consisting of a sac-like body about 1/16th mm. long containing undifferentiated reproductive tissue. The larva attaches itself to the host (a crab) and this tissue is forced through a special hollow process into the body of the host, to form a growing mass of cells which feed on the tissue of the host. We are not concerned with the rest of the life history. The parts of the larva bear a functional relationship to each other and are characteristic of the organism which produces them. We can ask what each part is for.

A remarkable parallel to this crustacean is to be found at a molecular level of organisation, that of the bacteriophage, a virus parasitic on bacteria. The organism in this case is a sac about 1/10,000 mm. long containing reproductive substance, desoxyribonucleic acid, with access to a tube which is attached by

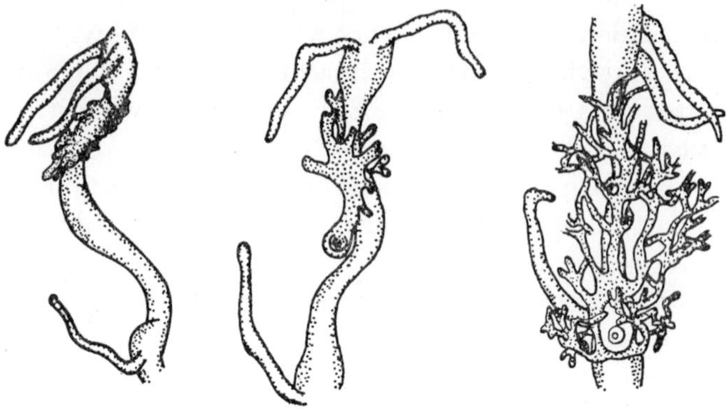

Figure 8. The larval stages of the parasitic crustacean *Sacculina*; showing the growth of the parasite internally along the intestine of the crab. (From J. A. C. Nicol, 1960. *The Biology of Marine Animals*, p. 603, after G. Smith. London: Pitman.)

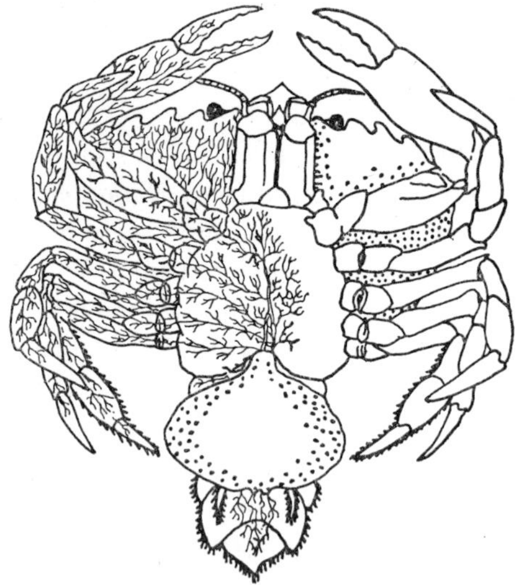

Figure 9. Diagrammatic representation of *Sacculina carcini*, showing the extensive system of roots which ramify through the tissues of the crab. (From Nicol, p. 602.)

specialised threads to the surface of the bacterium. The nucleic acid is forced through the tube into the body of the bacterium, where it multiplies at the expense of the bacterial substance. As machines the two parasites, the crustacean and the virus, are

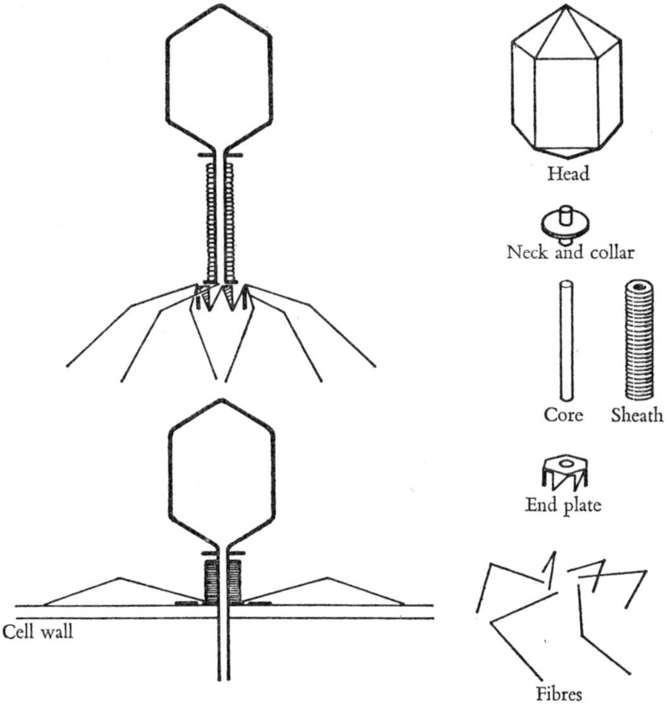

Figure 10. Structure of T4 bacteriophage based on electron micrographs. Magnification × 370,000. (Reproduced by permission of the World Publishing Company and Weidenfeld and Nicholson from *The Origin of Life* by J. D. Bernal. Copyright © 1967 by J. D. Bernal.) As contact with the surface of the bacterium is established the sheath shortens and the core passes through the bacterial wall, as seen in lower figure.

remarkably similar in spite of the enormous difference in scale of organisation.

But in the phage the parts show evidence of individual molecules of which they are composed. Whereas an isolated part of a large organism can give us unique information about the functional structure from which it is derived, the molecular functional parts can only give us ambiguous information. An isolated molecule of collagen might have been derived from any one of a variety of different structures or organisms. Precise ascription of a function to an *isolated* part begins to be difficult just as we approach the molecular level of organisation.

Nevertheless, these large molecules are in some respects intermediate between unique large functional parts and atoms which tell us nothing of the structure from whence they were derived. Thus, if we find molecules of a haemoglobin, the chemical class to which the red colouring matter of our blood belongs, we can infer with moderate probability that it was the respiratory pigment of a vertebrate animal. We may be wrong, for it may have been derived from other sources, where it had other functions. Thus it might have come from the roots of leguminous plants, where it is concerned with nitrogen fixation. But if all we know about the substance is that it is haemoglobin it is still moderately safe to wager that it was the respiratory pigment of a vertebrate animal.

Thus, when we examine living organisms, we find not only that their parts are functional, but that this characteristic fades and becomes equivocal at a molecular level. But, in addition, the illustrations given show that the same class of functional system can recur at different levels of organisation.

Another striking example of this is seen in certain slime-moulds. In species of the genus *Dictyostelium*, the organisms are small amoebae. In the feeding phase large numbers of these wander about exhibiting amoeboid movement, the pushing out of protoplasmic processes for locomotion and for the capture of food. Under certain conditions these amoebae aggregate together to form a compact sausage-like mass. In this genus, the amoebae retain their individuality; they become, as it were, 'cells' of the large mass. But the large mass itself as a whole then begins to move as a giant 'cellular' amoeba. Moreover it exhibits directed behaviour comparable to that seen in 'unicellular' amoebae. Thus we have amoeboid movement and behaviour both at a unicellular (or non-cellular) level and at a multicellular level: the

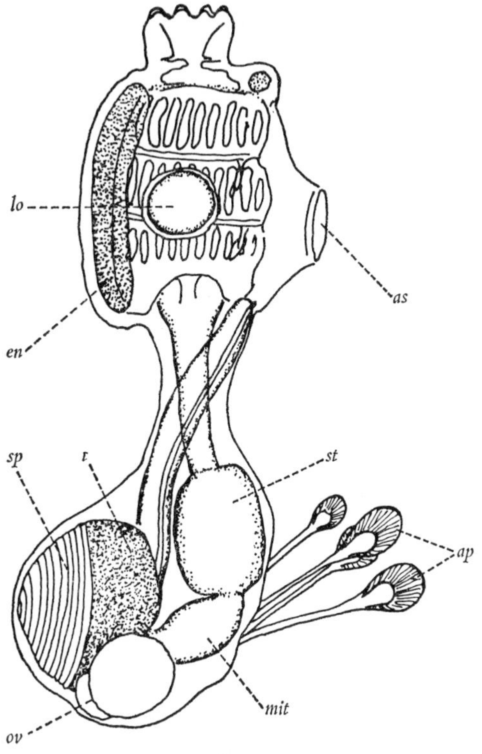

Fig. 11

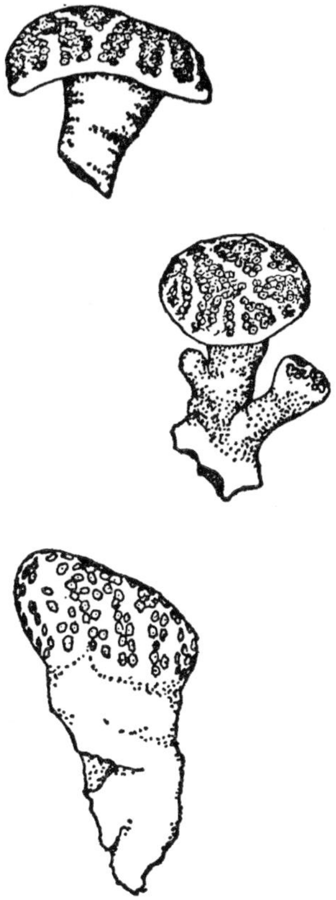

Fig. 12

Figure 11. A single organism of a sea-squirt *Trididemnum* (*Leptoclinum*) *tenerum*. *ap*, ampullae; *as*, atrial siphon; *en*, endostyle; *lo*, lateral organ; *mit*, mid intestine; *ov*, ovary; *sp*, sperm duct; *st*, stomach; *t*, testis. (From N. J. Berrill, 1950, *The Tunicata*, p. 117. London: Ray Society Monographs.)

Figure 12. Form of a colony of sea-squirts related to *Leptoclinum* (after W. G. van Name, 1945, *Bull. Amer. Mus. Nat. Hist.* **84**, 148). The figure shows (middle) the entire colony of *Distaplia bursata* and (above and below) different parts of it. Although, as can be seen from Fig. 11, *Leptoclinum* when adult is an elaborate organism, nevertheless, when these are united in a colony and form, as they do in this species, encrusting masses over seaweed, etc., the colony as a whole in its 'migrations' shows something very like simple amoeboid movement.

same class of phenomenon recurring at different levels. In certain cases something very like amoeboid movement is seen at a still higher level in the locomotion of colonies of cellular animals—as in the migrations of colonies of the sea-squirt *Leptoclinum*.

But, again, the recurrence of a dynamic system of a given class is not confined to the functional systems of living organisms. An eddy in a teacup, a whirlwind, a sunspot and even a galaxy have at least certain features which place them in a common class. In the phenomena both of amoeboid movement and of vortices in the physical world, we can of course differentiate sub-classes which meet the specification of the class in different ways. The peculiar feature of such classes in living systems is the amount of detail in the specifications of the general class and the significance we attach to it, even though the specification may be met in very different ways. Thus animals exhibit the power to learn, but we must be careful about asking, 'What is the mechanism of learning?', because the specification of learning may be met in very different ways—and even in the same organism.

It is clear that our general discussion today of animate and inanimate systems fails to provide a sharp distinction which would enable us to make the everyday statement, 'This creature is alive'. In the next lecture we must therefore consider the nature of living matter.[13, 14]

SUMMARY

1. In practice, the scientist, like the man in the street, accepts the reality of his perception of an external world of real objects. Conviction of this is based on the mutual consistency of the relations of all our impressions about what we take to be the natural world, however far the consequences of those relations are followed. These relations are directly perceived.

2. When we examine this 'real world' it is built up of objects that endure, though they decay, and dynamic equilibria. In addition there are cyclic phenomena. The atmosphere provides a striking example of such an equilibrium which has endured for an immense period of time.

3. Our detection of enduring objects and equilibria is subject to the limitations of our 'sensory spectrum'. We do not become directly aware of phenomena which are too small, too large, too brief, or too slow.

4. Matter as we perceive it in the solid, liquid and even the gaseous states assumes remarkably complex and ordered patterns—the enduring objects and the dynamic equilibria. The paths by which they assume these patterns are governed by the laws of thermodynamics, of which the second law—that which states that disorder tends to increase—is the most important, together with Le Chatelier's principle of least action.

5. The highly ordered structure of living things is attained without contravention of the laws of thermodynamics. That some complex systems can attain a high degree of locally ordered structure, at the expense of a much greater increase of disorder elsewhere in the system, is a feature of living things. But it is not restricted to these. Thunderstorms and other complex physical phenomena may share these features with living organisms.

6. A preliminary consideration of the characters of living things shows other features, including the power of growth and of reproduction, which can be found also in the inorganic world.

7. Living things, however, are characterised by the possession of organisation into existentially dependent parts. These parts have a 'purposive' character, and this character enables the scientist to make useful predictions about them which can be verified. But the same thing can be said of machines.

8. In living organisms there is a limit at the molecular level at which the 'purposive' character of an existentially dependent part becomes equivocal.

9. It is a characteristic of living things that they can exhibit organisation at many levels simultaneously: as in cells, organisms, colonies, and so on. The same classes of phenomena can be found to recur at different levels of organisation. But this also is true of machines and some complex physical systems.

10. The sharp distinction we commonly draw between living and inanimate objects arises in part from the special attention we

pay to physical systems simple enough for easy analysis and logical treatment.

When we consider the whole range of physical phenomena we often perceive features we consider as characteristic of life. Nevertheless, living creatures comprise the most striking class of object presented to our everyday experience. We must therefore consider further in what their special character rests.

3

Living systems and natural selection

OF ALL THE OBJECTS we perceive, none are more various and striking components of our world than living creatures: man, animals and plants.

Whatever may be said about the difficulty of defining life and of analysing life for distinguishing characters, our native power of separating the living from the inanimate depends, as it has always depended, on three things. These are to be recognised in plants but are most evident in animals. They are: first, what organisms *do* is different from what *happens* to stones and other inanimates. In common speech we use final causes to describe the behaviour of animals. Second, the parts of organisms are functional. Again, in common speech we use final causes to describe what a limb is for. That is less evident in plants than in animals because we cannot use the evident analogy with the functional parts of our own body, and also because so much of a plant's adaptive structure is beyond our sensory spectrum at the cellular and chemical level demanded by photosynthetic machinery, a difference that is at the root of the contrast between the syllabuses of lectures in botany and zoology. But the very use I have just made metaphorically of the word 'root' illustrates our recognition of functional characters in plants. The third native feature in our recognition of organisms is the obvious difference of their material substance. Meat, liver and the petals of flowers seem quite unlike earth, air, fire and water.

In former times, many attempts were made to connect the peculiar features of living organisms, which I have listed, with the special properties of organic matter. After Wöhler's synthesis of urea and the steady progress of organic chemistry during the nineteenth century it became clear that organic chemistry was the chemistry of carbon, and that though this was certainly a strange element with remarkable powers of combination it demanded no special laws inapplicable to inanimate matter. Whilst organic molecules were the essential material substance of living creatures, it was not their molecular chemistry which seemed to differentiate the living from the non-living.

William Whewell in his *Philosophy of the Inductive Sciences* of 1847[1] still invoked 'vital forces' in distinguishing living matter. What Whewell meant is shown in his discussion of nervous action:

> Vital endowments incapable of being expressed or explained by any comparison with mechanical, chemical or electrical forces...

Twenty-one years later Thomas Henry Huxley[2] gave his contemporary mechanistic view of the physical basis of life. He notes the emergence of new properties when the chemical elements combine to form compounds. The properties of the compound *water* do not require the assumption

> that a something called 'aquosity' entered into and took possession of oxidated hydrogen as soon as it was formed...What better philosophical status has 'vitality' than 'aquosity'?

and later:

> The elements of living matter are identical with those of mineral bodies; and the fundamental laws of matter and motion apply as much to living as to mineral matter; but every living body is, as it were, a complicated piece of mechanism which 'goes', or lives, only under certain conditions.

These statements are very much to the point—provided that we recognise clearly that we are drawing a bill on the future success of natural science in providing explanations of the emergent properties of higher levels of organisation in terms of lower units. What disarms, not to say disheartens, the careful logical critic is that nature so often honours the post-dated bill so brashly presented by the mechanist.

For it is fair to ask the mechanist, 'Just how do you think this comes about?' As Lord Kelvin[3] said:

> It seems to me that the test of 'Do we or do we not understand a particular point in physics?' is 'Can we make a mechanical model of it?'.

I know that respect for mechanical models has faded in these days in preference for mathematical ones. But I would now only remind you that the mechanical model with all its defects has one virtue that is its own: it guarantees that a system of the class envisaged really works, and that at least in that respect we have not unwittingly let by an assumption which vitiates analogy. So may we not ask the mechanist, 'What is your model?'.

The question is a searching one and all too easily glossed over by the optimistic mechanist, to give the impression that all is sound in the edifice of his interpretation of the natural world, whereas that may only be held together by papering over the missing joints with by-and-large-isms. But that does not mean that the mechanist is forever incapable of providing answers to his questioners. It is instructive to consider the problem which Clerk Maxwell set before the physiologist in his article on the 'Atom' in the *Encyclopaedia Britannica* for 1875. The question he is asking is: Just how does the unique complexity of the individual arise from the minute egg?

A cube whose side is the 4000th of a millimetre may be taken as the minimum visible for observers of the present day. Such a cube would contain from 60 to 100 million molecules of oxygen or of nitrogen; but since the molecules of organised substances contain on an average about 50 of the more elementary atoms, we may assume that the smallest organised particle visible under the microscope contains about two million molecules of organic matter. At least half of every living organism consists of water, so that the smallest living being visible under the microscope does not contain more than about a million organic molecules. Some exceedingly simple organism may be supposed built up of not more than a million similar molecules. It is impossible, however, to conceive so small a number sufficient to form a being furnished with a whole system of specialised organs.

Thus molecular science sets us face to face with physiological theories. It forbids the physiologist from imagining that structural details of infinitely small dimensions can furnish an explanation of the infinite variety which exists in the properties and functions of the most minute organisms.

A microscopic germ is, we know, capable of developing into a highly organised animal. Another germ, equally microscopic, becomes, when developed, an animal of a totally different kind. Do all the differences, infinite in number, which distinguish the one animal from the other, arise each from some difference in the structure of the respective germs? Even if we admit this as possible, we shall be called upon by the advocates of Charles Darwin's theory of Pangenesis to admit still greater marvels.

His question is a rhetorical one; a question expecting the answer 'No'; and he goes on to remind us that

...one material system can differ from another only in the configuration and motion which it has at a given instant.

His unstated implication is that material configuration alone does not suffice to explain the unfolding of the adult organism.

Clerk Maxwell's[4] problems are the essential ones of development and of heredity. In the modern phrase, how is the detailed programme of a unique and complex living organism *coded* in the minute germ from which it springs? And how are we to conceive this coded information to be transmitted from one generation to another? How in fact are the molecules arranged, and what do they do? One of his difficulties arises from his implicit assumption that for an adequate mechanical model the unique and complex multimolecular structures in the adult must be coded by structures in the germ built up of quite small molecules—of which in fact there are far too few to do what is wanted.

It would not at first sight have helped him to say that in fact there were far fewer molecules in the germ even than he supposed, and that these molecules were several thousand times larger and longer than those with which the chemist was then acquainted; still less could he have suggested that the code was not a code of gross molecular arrangement such as we see in the structure of an adult creature, but was a configuration of parts within certain molecules. Yet we now have strong evidence that this is in fact the case.

Clerk Maxwell's impasse was unreal. As we look back we can see this. It is so easy to accept a conclusion once we have been shown how to arrive at it. Had Clerk Maxwell in 1875 suggested that these things were coded in gigantic molecules, his would have been speculation as rank as any in Robert Chambers's *Vestiges of Creation*.[5]

Difficulties of another sort are to be seen in Karl Pearson's discussion on 'Mechanism and Metaphysics in Theories of Heredity'. His *Grammar of Science*[6] published in 1892 had a profound effect on the mechanistic development of biology that followed in this century. Rather surprisingly we find him taking to metaphysical task none other than Weismann,[7] the author of that theory of heredity which was the forerunner of modern genetics. Of him, Pearson says:

This theory is summed up in the formula of the 'continuity of the germ-plasm'. According to this theory there exists a substance of a *definite chemical and molecular structure* termed germ-plasm which resides somewhere in the germ-cells, from which reproduction takes place. In each reproduction a part of the germ-plasm 'contained in the parent egg-cell is not used up in the construction of the body of the offspring, but is reserved unchanged for the formation of the germ-cells of the following generation'...Now this hypothesis of Weismann as a conceptual mode of describing our perceptual experience seems to be of considerable value, but the author weakens his position throughout by projecting his conceptions into the phenomenal world, where up to the present nothing has been identified as the perceptual equivalent of germ-plasm. It is this transition from science as a conceptual description of the sequences of sense impressions to metaphysics as a discussion of the imperceptible substrata of sense-impressions which mars biological as well as physical literature.

Perhaps. But Weismann's bill on the future has in fact been dramatically honoured. For the poor scientist, 'is there anything better, anything better? Tell us it then!'.

The past history of ideas on the nature of life need not deter us from trying to envisage models of its distinguishing features. Consider the power of reproduction. Whilst diligent search may bring to light examples of reproduction and growth amongst inanimate systems, these are almost universally characteristic of living things. Their 'purposiveness' is indeed the most striking and essential character of them;[8] and though the purposes disclosed are very varied all can in some degree be related to the maintenance and reproduction of the species or group of individuals to which the organisms belong. This purposiveness is not only to be seen in the functional character of the parts, but also in the life history and, particularly among animals, in their goal-directed behaviour. The adaptations of the successive stages of a parasitic worm towards ensuring the future transmission of some individuals, at least, from host to host are as astonishing as they are economically important.

I must make it clear that I am using the word 'purposive' here as the label in everyday use for structures that seem directed to assignable ends. In everyday use the word carries with it overtones of association with volition in ourselves. That difficulty could be got over by choosing a special word, such for instance as 'proleptic' for such phenomena. The danger here is that if I choose a term which others have used it already possesses overtones of association, whilst if I choose an entirely new term I place a barrier between myself and apprehension by an audience. One must stop and think: 'Oh yes, that's what in everyday life we call purposive.' There is danger in both ways. To adhere to the common word carries the danger of its secondary associations. To invent a new one risks the presence of other associations of which we may not even be conscious; and worse still we may unwittingly exclude some aspect of the phenomena included in common recognition; and we may set up a barrier of language in communication with others. Therefore I prefer, where possible, to use an everyday word like 'purposive', with this warning, that it is to be used as an empirical label and is not to be allowed to carry implications of the special origin of its characteristics.

I would make this clear by an example. As Charles Darwin saw, the amazing structural modifications of plants to ensure fertilisation by insects are amongst the most striking examples of apparent design: that is, they have something in common with the quality of purposive ingenuity which we see in humanly designed instruments. It does not follow at once from this that the origin of the purposive quality of these is the same. All that we can say at the moment is that both the parts of living organisms and the parts of man-made machines conform to a certain specification: that when we examine an isolated part we can usefully discuss its presumptive future use. We can make verifiable predictions on that basis. Without further examination we must not at once jump to the conclusion, as Ray and Paley did, that all 'tenders' (if I may borrow an expression from industry) to meet this specification are of the same class.

To return to the power of reproduction, though this is not a quality by which we immediately recognise a living organism, it is certainly a distinctive feature to be seen in living things on investigation. I have already remarked the appearance of reproduction in inanimate systems, particularly in vortex rings in water. But reproduction in these does show important differences from reproduction in living things. As the vortex ring I described in my last lecture descends through water, it generates new vortex rings, but we do not find that each ring truly reduplicates itself. In the experiment, the drop of fluid of slightly higher density generates a vortex ring when it drops into water. The ring expands and slows down through friction. Observation shows that its revolution is finally brought to a halt. We now have a ring of the denser fluid within the water; but it is no longer a vortex ring. From various points round the circumference of this ring the denser fluid starts to drop downwards and at each of these points a fresh vortex ring is formed. This in turn slows down and the cycle is repeated. Thus each vortex ring does not directly produce new rings like itself; it is rather that it creates a situation in which new rings can, so to speak, be 'spontaneously' generated. There is reproduction, but not reduplication.

This kind of reproduction differs strikingly from that of living organisms. In these reproduction is essentially a reduplication. It is true that a parent amoeba by dividing into two loses its single existence, but at no point during the process does there cease to be an organism. Moreover, the reduplication is not merely that of the general class of system to which the parent belonged, but involves a reduplication of characteristics of the individual parent or parents.

How does this power of reduplication relate to what is known of the structure of living organisms? With regret I must now embark on a fairly technical description of the living cell and living substance. My regret is not lessened because I am not a biochemist but an ordinary general biologist and what I have to describe concerns scientific work which, in the hands of many able specialist investigators, is advancing more rapidly than any other part of natural knowledge except nuclear physics and the principles of nervous action. All the same, if we are talking about the relationships of the sciences, I must try to display as clearly as I can what I believe we are trying to talk about.

As we see them today, living creatures—plants, animals and bacteria—are objects composed of organic matter of which we hold two chemical substances, protein and nucleic acid, to be particularly important, together with the high percentage of water. Notwithstanding the fluidity of much of their substance, solids, particularly in the form of fibres, play an essential structural part.

The bodies of the higher plants and higher animals are built up of cells: muscle cells, nerve cells, blood cells. But whether cellular or not, the bodies of living creatures resemble complex machines composed of parts to which functions can be attributed and to which a purpose for use in the future can be assigned. Like complex machines, there seems an evident improbability about them in the sense that the chance that a lot of carbon, hydrogen, nitrogen, oxygen and other molecules would jump into place to create such an elaborately interdigitated functional machine as a man seems indefinitely remote.[9]

So far the chief differences of a living system from a growing physical system like a thunderstorm are the different materials of which it is composed, the presence of solids, and the very much greater elaboration of the parts. Whereas the vertical current of a thunderstorm just begins to merit the functional description of 'what it is for', such a description is absolutely characteristic of the parts of animals.

The bodies of the higher animals are composed of tissues, muscle, nerve, digestive glands. These in turn are composed of cells: living units with nucleus and cytoplasm, and bounded by a cell membrane which limits and controls exchange with the external fluids of the body and of the outer environment. The cells are semi-independent individuals. They can be cultivated in a test-tube, where they will grow and reproduce independently.

But, assembled in the body, they compose the organism, which is thus the result of the configuration of these simpler individuals with their secretions. Thus cells are the smallest units in the higher organisms which can be separated and can maintain themselves as respiring, growing and reproducing objects. In creatures like Amoeba there is no division into cells. The whole individual can be compared to a single cell.

Just as in the grosser parts of an animal, examination shows that even the cells themselves are built up of functional parts: nuclei, self-reproducing bodies known as mitochondria and plastids, and so on. I am not now concerned with precisely what goes on in this varied and elaborate machinery. But in contrast with what happens to organic food materials when subjected to chemical agents in test-tubes, or even when they are subjected to the ground-up residue of living cells, their fate within the living cell shows clear evidence of a highly ordered sequence of chemical processes. A diagram of the chemical events which take place in a cell recalls the diagram of the events which take place on the production line of a factory. In both cases the ordering of the events implies the existence of spatially and temporally ordered machinery. It is quite clear that the parts of cells have functional significance; we can usefully ask, 'What is a nucleus or a mitochondrion for?', as an experimental directive—in the same way as we can ask this about a limb or a sense organ.

This is one of the most remarkable advances in scientific knowledge. Since 1859 we had been accustomed to assign the purposive structure of gross organs to the operation of natural selection. But almost nothing could be said of the origin of the structure of the cells of which these gross organs are composed. Cells were said to contain a nucleus and a vital substance, protoplasm, of somewhat mystical properties. All that is now gone, and purposive structures are now evident within the cell itself—just as they are even in the parasitic bacteriophage virus. And the modern study of the ultrastructure of these units describes this in terms of molecular configurations. The passage from comparative anatomy to molecular structure has in fact been effected.

As in the bacteriophage, the purposive character of these cell structures becomes equivocal at the molecular level. But so long as we can assign function to these minute structures, the inductive argument that such coordinated ultrastructure of the cell must arise through the operation of natural selection can hold, just as it has done for gross adaptive structures since 1859.

One most important feature of this cellular machinery both in its structure and in the chemical events which take place in it is its universality. Notwithstanding the enormous evolutionary divergence between a bacterium, an oak-tree, and a man, there is a remarkable community of ultrastructural parts, and the same chemical operations can be detected again and again in totally different organisms, and they are used for various ends. Repeatedly we find in all sorts of organisms that the energy of living processes is supplied by the same particular energy-rich molecule, adenyl-triphosphoric acid.

In the minute structures of the cell we begin to meet similarities which are not, or not necessarily, of evolutionary origin, but which arise from the severe limitations of the available kinds of atoms and molecules and their properties. Organic material is based upon the chemistry of carbon. This is no evolutionary accident, such that on some other planet we might expect selenium, iodine or tantalum to have been selected to fill that role. Carbon is unique. The kinds of organic molecule with the unique properties needed for living machinery are unique. It is conceivable that in other worlds and other environments organisms could exist built from quite a different kind of chemical machinery. For various reasons that seems unlikely—but even if it be true, the rigid limitations of the properties of matter would demand that only a limited number of classes of such machinery could exist. The presence of carbon in two organisms gives no evidence of their evolutionary relationship; nor, necessarily, does the presence of particular molecules which happen to have valuable constructional properties for a living machine.

The most remarkable example of this community of chemical organisation is the presence in every living organism that can

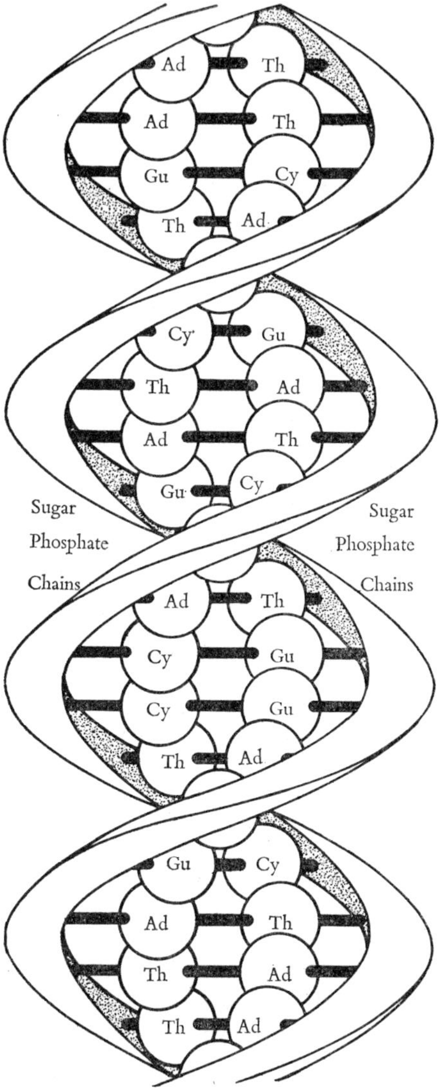

Figure 13. The double helical form of DNA. A schematic representation. (From Bruce Wallace, 1966, *Chromosomes, Giant Molecules and Evolution*. New York: W. W. Norton & Company, Inc. Figure adapted from Maurice Sussman, *Growth and Development*, 2nd © 1964. Reproduced by permission of Prentice-Hall, Inc.)

unequivocally be so called of a particular kind of highly complex substance, desoxyribonucleic acid, DNA. It is chemically unique, and it so happens that this molecule itself provides a unique and intricate piece of machinery.[10] It is far the most remarkable example of that peculiar fitness of properties of matter for the existence of living matter to which Laurence Henderson[11] drew attention so many years ago.

The study of Mendelian inheritance has made it abundantly clear that the unique individual characteristics of living organisms can be shown in the vast majority of cases tested to be determined by unit characters, the genes, which are inherited unchanged. Breeding experiments show that these character-bearing genes can be arranged in serial orders—like beads on a string. In the higher organisms a system which corresponds in structure and behaviour precisely to the material 'model' which we predict from the results of breeding experiments is to be found in the thread-like chromosomes of the nucleus. DNA, together with protein, is always present in these chromosomes. There is good evidence for the existence of essentially similar machinery in bacteria and other simple creatures.

As we have said, it is manifest that living organisms can reduplicate the features of their kind. Study has shown that an essential feature of the reduplication of organisms is the reduplication of chromosomes. This reduplication can now, in turn, with high probability be assigned to the reduplication of the particular kinds of DNA molecule characteristic of the organism. The position is well summarised in an essay by Dr Broda:[12]

> It was only about 1930 that people started working on the chemical basis of inheritance, and only since the war that nucleoprotein, and more particularly desoxyribonucleic acid (DNA), has been established as the carrier of heritable 'information'. The evidence is sparse, yet it is overwhelming. The line of argument may be presented thus: metabolic processes can eventually be traced to enzymes, which must themselves be synthesised but cannot duplicate themselves. The proteins of an organism are specific and this specificity is heritable; heredity is carried on the chromosomes, in which DNA is always present. DNA is in general not found elsewhere in the cell.

The serially particulate character of inheritance is that correlated with the serially particulate character of the arrangement of a molecule of a particular sort. By implication from the results of genetical studies, these molecules must have two powers: (1) on some occasions they must reproduce other molecules of precisely the same sort and in the same ordered arrangement, and (2) on

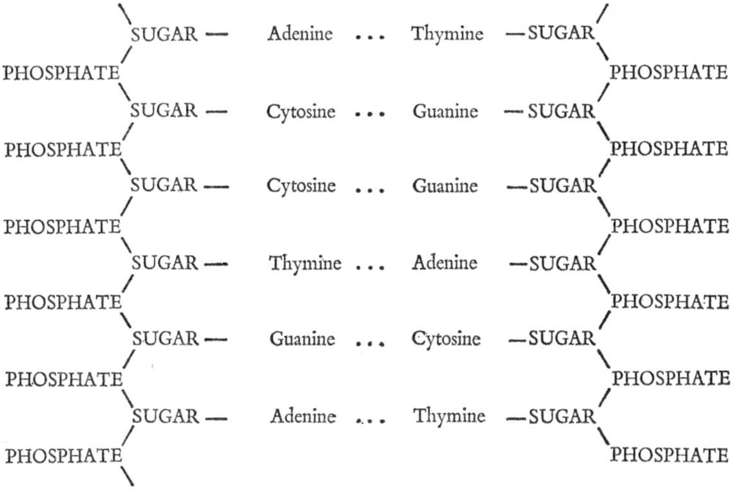

Figure 14. DNA. Diagram to show the principles of linkage between two chains to form a two-stranded structure. (From A. R. Peacocke, 1965, p. 28, *Biology and Personality*, edited I. T. Ramsey. Oxford: Blackwell.)

other occasions they must set in motion the ordered production of substances which will determine the successive production of materials and the successive direction of the metabolic processes of the cell so as to synthesise another living individual of the same kind as the parent.

The sequence of operations which results in the formation of the individual is thus supposed, and with great probability, to be held as a 'code' in the ordered DNA molecules. It has been one of the triumphs of molecular biology that Watson and Crick on the basis of X-ray and crystallographic data were able to prepare a molecular model which has already repeatedly been shown to be consistent with the known chemical, cytological and genetical requirements of the reduplicating system.

As a molecule, the nucleic acid DNA is of enormous length. It consists of a chain of parts, the so-called nucleotides. There are in fact only four sorts of these nucleotides, and in each sort of DNA molecule these are arranged in a different serial order, just as four sorts of beads can be arranged along a string in an enormous variety of serial orders. Such a serially ordered uniqueness can be compared to a linear code (Fig. 17) bearing information—a fulfilment of Weismann's prediction.

In Watson and Crick's model, the DNA strings occur in pairs, arranged side by side along their length. Indeed the double thread or helix (for the two threads are twisted round each other corkscrew-wise) can be considered as a minute elongated aperiodic crystal (Figs. 13 and 14). That the two DNA threads can be held together in this way is possible because the four kinds of nucleotide beads of which they are composed are not identical, but happen to exist as two sets of complementary parts. The succession along one thread resembles the projections and hollows of the edge of a key; and the succession along the companion thread resembles the wards of a lock (Fig. 15). When reduplication occurs, each of the DNA strings separates from the other, and each is able to assemble a fresh replica of its companion by the collection in the right place and order of new beads of nucleotide units, from the dissolved organic matter in the surroundings.

What happens in effect is very like the successive reproduction in a factory of castings from moulds and moulds from castings. There is, however, this very important difference. Moulds and castings gradually become defaced and worn out by repeated use. The molecular moulds or templates of nucleic acids are composed of atoms, the individuals of each kind of which are indistinguishably alike and cannot wear out. Unless the molecular template is at some time broken it can never wear out and precisely similar castings and moulds could be produced from it for ever. It is the fact that the organisation of living creatures, however large and gross their structure, is ultimately determined by a molecular, and therefore precisely repeatable, template which distinguishes them from other macrophysical systems like thunderstorms.

The reduplication of nucleic acid chains is of course only the first, though essential, part of the machinery of reduplication of a whole organism. In the living cell, the DNA templates not only have the power to reduplicate themselves, but to manufacture

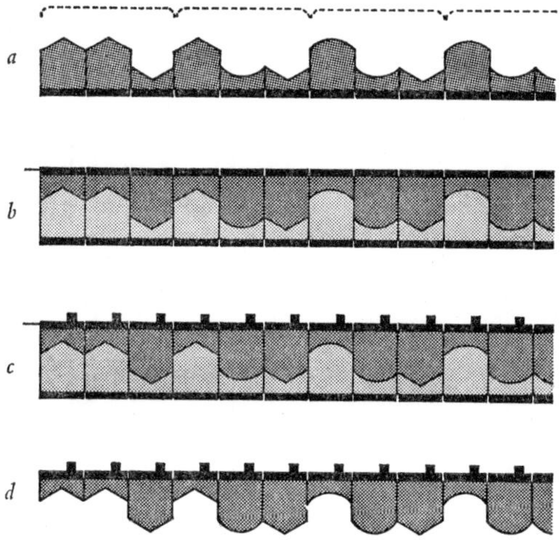

Figure 15. Synthesis of DNA begins with a single strand of the normally two-stranded molecule (*a*); nucleotides then assemble themselves along this strand to reconstitute the other (*b*). Presumably by a similar process (*c*) a strand of DNA can assemble a strand of RNA (*d*). The knobs along the RNA backbone represent hydroxyl (OH) groups. (From *Scient. Amer.*, December 1959, p. 58.)

other components of the cell such as a strand of RNA (containing ribose instead of deoxyribose) which, while not able to reproduce itself can carry the information as to the particular sequence of amino-acids. This, by a series of steps, can serve as a template which by the action of other substances such as enzymes (catalyst proteins) can ensure the synthesis of the corresponding protein. Fig. 16 shows the first established amino-acid sequence for an enzyme. These operations again require a supply of energy which is derived from an energy-rich molecule, adenyltriphosphoric acid, which is itself a nucleotide combined with further phosphoric acid.

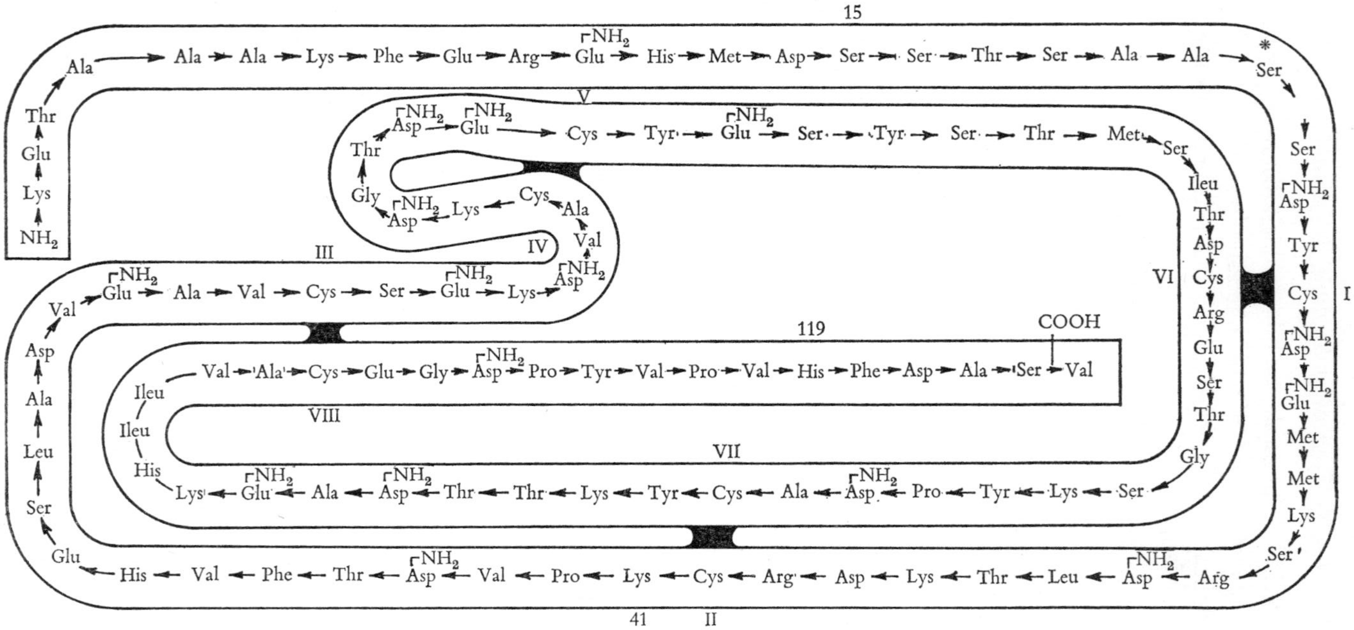

Figure 16. The amino acid sequence of the enzyme ribonuclease from bovine pancreas. (From D. H. Spackman, W. H. Stein and S. Moore, 1960, *J. Biol. Chem.* **235**, 648; and D. G. Smyth, W. H. Stein and S. Moore, 1962, *J. Biol. Chem.* **237**, 1845.)

The production of these various substances takes place in an ordered manner in time and position, to make the cells, the tissues, and the organism. All this sequence is coded in the original pattern

2nd→ 1st ↓	U	C	A	G	3rd ↓
U	Phe	Ser	Tyr	Cys	U
	Phe	Ser	Tyr	Cys	C
	Leu	Ser	ochre	?	A
	Leu	Ser	amber	Tryp	G
C	Leu	Pro	His	Arg	U
	Leu	Pro	His	Arg	C
	Leu	Pro	GluN	Arg	A
	Leu	Pro	GluN	Arg	G
A	Ileu	Thr	AspN	Ser	U
	Ileu	Thr	AspN	Ser	C
	Ileu	Thr	Lys	Arg	A
	Met	Thr	Lys	Arg	G
G	Val	Ala	Asp	Gly	U
	Val	Ala	Asp	Gly	C
	Val	Ala	Glu	Gly	A
	Val	Ala	Glu	Gly	G

Figure 17. The genetic code today. The four standard bases, uracil, cytosine, adenine and guanine, are represented by the letters U, C, A and G respectively. The code is known to consist of three-letter 'words', the triplets (e.g. GAC, UUA etc.) each of which is the code for an individual amino-acid. The first base of any triplet is indicated on the left, the second base at the top, and the third at the right of the figure. The twenty amino acids are represented by their standard abbreviations, thus: PHE stands for phenylalanine, etc. The triplets marked 'ochre' and 'amber' are believed to signal the termination of the polypeptide chain. Those associated with chain initiation are not marked in this figure. (After F. H. C. Crick, *Proc. Roy. Soc.* B, **167**, 334, being the Croonian Lecture 1966: The Genetic Code.)

of the DNA (Fig. 17). We can make an analogy with the ordered sequence of events in a game of chess, from the opening gambit with a configuration of individual pieces to the complex pattern of functional implications in the end game.

We may consider it remarkable that material units in certain configuration possess the properties of a reproducing organism, or, for that matter, those of a machine, like Edsac.[13] But the material world is such that this is so; why that is, is a second question. But to consider the configuration of the DNA molecule itself. It is a unique structure of ordered parts. This order confers on it the following property: when in the appropriate environment—of 'food' and 'energy'[14]—a sequence of events takes place which results in the formation of an organism. Is the special reduplicating feature of the DNA molecule with all its consequences a reflection or modification of the nucleotide units of which it is composed consequent upon their organisation into that particular sort of DNA molecule? We are accustomed today to realise that no particle of matter, nor any event, is wholly uninfluenced by the rest of the natural world. The concept indeed goes back to one who was an experimental physicist and not pre-eminently a mathematician, Michael Faraday.[15] In 1844 he made an attack upon Dalton's atomic theory. He asks: What do we know of an atom apart from its force?

He says:[16]

> This view of the constitution of matter would seem to involve necessarily the conclusion that matter fills all space, or at least all space to which gravitation extends; for gravitation is a property of matter dependent on a certain force. In that view matter is not merely mutually penetrable, but each atom extends, so to say, throughout the whole solar system, yet always retaining its own centre of force.

Are we to seek the special properties of the DNA molecule as an organised whole in the mutual influence of its parts after Faraday's fashion? That seems to me to come very near to a conclusion of Whitehead's[17] about this problem. In discussing the living organism he says:

> The prompt self-preservative actions of living bodies, and our experience of the physical actions of our bodies following the determinations of will, suggest the modification of molecules in the body as the result of the total pattern. It seems possible that there may be physical laws expressing the modification of the ultimate basic organisms when they form part of higher organisms with

adequate compactness of pattern. It would, however, be entirely in consonance with the empirically observed actions of environments, if the direct effects of aspects as between the whole body and its parts were negligible. We should expect transmission. In this way the modification of total pattern would transmit itself by means of a series of modifications of a descending series of parts, so that finally the modification of the cell changes its aspect in the molecule, thus effecting a corresponding alteration in the molecule—or in some subtler entity. Thus the question for physiology is the question of the physics of molecules in the cells of different characters.

But in all this we must be careful not to overshoot the mark and start to look for the origin of the special features at too low a level. A petrol engine can be considered as an organism. Its functional unity depends upon the configuration of its parts. When the cylinder-head is finally bolted on we do not find that all the rest of the parts change their character, so that the whole can become a petrol engine. Similarly in the DNA molecules it is the configuration that confers the unique property, and not a subtle influence of the whole upon the nature of the parts. At a still higher level the same kind of thing can be said of the nervous machinery of the brain. Its remarkable properties are related to the pattern of its cellular organisation. They are not to be sought in some influence of the whole upon the molecules which compose the cells. Just as in computing machines, the remarkable emergent properties of the 'organism' are to be sought at a comparatively gross level.[18]

In the whole material world, molecules, men and mountains, we can perceive an enormous set of configurations at different levels of complexity. Some, as in organisms and machines, have very remarkable properties at some level of organisation. Such special properties have been referred to as 'emergent'. The notion was originally put forward by G. H. Lewes (1875)[19] in discussing the relation of a whole to its parts:

All quantitative relations are componental; all qualitative relations elemental. The combinations of the first issue in Resultants, which may be analytically displayed; the combinations of the other issue in Emergents which cannot be seen in the elements or deduced from them. A number is seen to be the sum of its units; a direction of movement is seen to be the line which would be occupied by the body if each of the incident forces had successively acted on it during an infinitesimal time; but a chemical or vital product is a combination of elements which cannot be seen in the elements. It emerges from them as a new phenomenon.

According to Lloyd Morgan[20] the appearance of matter, of life and of mind are to be considered as emergent phenomena of this sort. As an expression of the empirical fact that at various grades of organisation material configurations display new and unexpected phenomena, and that these include the most striking features of adaptive machinery, this is very much to the point. Whether in fact we can prove in any or all of such cases that the assumptions needed to describe a simpler grade of organisation must necessarily require additional ones to describe the higher grade, that we must add to the assumptions needed to describe all we know about atoms if we are to extend our description to molecules, is more uncertain.[21] The assumptions are required not for the objects we perceive but for those aspects of them that we try to represent by our contemporary models. The difference between living and inanimate matter is not the same for us as for Lewes in 1875. Our knowledge of molecular structure does not make it self-evident that in passing from an atom to a molecule, or even to a replicating molecule of DNA, we must add novel assumptions to account for what we perceive. We perhaps cannot do without such assumptions now, but we may become wiser in the future. But so long as we treat as 'emergent' phenomena associated with organisation which are not evident in separate parts, whether or not the assumptions needed to describe the separate parts could of themselves lead us to the special properties of the organised system, the notion is of value.

Systems with the complex emergent properties of organisms are few amongst the whole set of possible systems. It is the essential feature of the theory of Darwinian evolution that these unique systems have been reached through the action of natural selection. Under the stress of natural selection acting on inherited variation, species can evolve along a path of systems, always functional but passing to related systems with new functions during evolution.

For the evolution of adaptive structure the mere power to reduplicate is by itself of little value. The situation is, however, wholly different if the reduplicating system can undergo variation, which can then in turn be inherited by reduplication. In simple, spontaneously generated systems, natural selection can do no more than allow one individual to grow at the expense of others by annexing to itself the common food supply. But as Charles Darwin showed, if systems can vary and the variation is heritable, then offspring adapted to a changing environment will be selected and any structural modifications which would lead to greater adaptation will be preserved.

Such heritable variation can be detected in Mendelian inheritance, in the chromosomes, and in their DNA constitution. X-rays or other agents can modify an element in the nucleoprotein chain; even the detachment of a single electron from a gene may cause a heritable mutant. Moreover, changes of this sort may also occur 'spontaneously' with the indeterminism of quantum changes.

In this cursive review of natural objects, from the simplest ones to the most complex examples of living machinery, we have come a very long way. It is the purposive character of their structure and behaviour which particularly distinguishes living organisms. That is a character which they share in a great degree with the complex machines we make; and though these follow our human designs their purpose is not a peculiarity of the kind of materials of which they are composed. Since Charles Darwin's demonstration that natural selection acting on inherited variation must lead to purposive adaptation in the structure of living things, such purpose could be considered as the negative image of blind chance. But natural selection is permissive rather than constructive. It might be that the possession of an eye would make the reproductive survival of a species more probable. But as I have tried to show in another essay, the possible form of an organ of vision is severely restricted by the properties of matter and the principles of engineering construction peculiar to this universe. There are so few ways in which an eye can be made. The principle of the photographic camera and of our own eyes is one of these ways.

The material parts of both men and machines can only be constructed on certain engineering principles; and only limited kinds of material are available to make them.

Starting with elementary particles and passing up through atoms, molecules, the remarkable states of dynamic equilibrium attained by matter in the mass, the variety of machines that can be made, the special reduplicating powers of DNA, and the whole variety of living organisms, we see the objects of the natural world as an immense array of varying complexity though nevertheless with severe limitations.[22] Natural selection has in fact enforced evolving species of organisms to traverse certain routes through that complex array of possible systems. We are still left with underlying qualities of this array. Considering the small number of elements available for the construction of organisms, it seems strange that such structures can in fact be built up. It is indeed a remarkable fact about the pattern of possible functional structures that routes through them, connected by their concomitant properties, exist.

All the same, whatever the complexity of possible objects at different levels of organisation, it is natural selection which ensures the survival of such systems, and we can see it to be one of the laws of thermodynamics which describes the fate of matter and energy in complex systems. But like the second law this only tells us the direction in which change is taking place. The classes of organisms which attain reality are selected from the classes of possible material configurations, the classes of events which are associated with them. The principles of their different systems of classification will be discussed in the next chapter.

SUMMARY

1. Whatever the difficulty of defining life, we in fact distinguish living from inanimate by three criteria: (a) what organisms *do* is different from what *happens* to stones; (b) parts of organisms are functional; (c) the material substances of the two are different.

2. In 1878 T. H. Huxley discounted emergent qualities as necessary in accounting for life. This view is provisionally approved

subject to the crucial proviso that such statements draw a bill on the *future* success of natural science in providing explanations of the emergent properties of higher levels of organisation in terms of lower units.

3. Clerk Maxwell's difficulties over the physical organisation of the germ cell have been overcome in a way which would have been inconceivable in 1875, namely, by the demonstration that coding is achieved by a few very large molecules, not a large number of small ones. Thus the bill was honoured in a manner quite unpredictable at the time it was drawn.

4. Reproduction in non-living and living systems is compared. In the former, as in vortex rings, new rings are *generated* but the new ones do not exactly *reduplicate* the old. In living organisms, on the other hand, reproduction is essentially a reduplication. It is the fact that the organisation of living creatures, however large and gross their structure, is ultimately determined by a molecular and therefore precisely repeatable template, which distinguishes them from large, purely physical systems like thunderstorms and vortex rings.

5. The 'bodies' of living organisms, whether they be single cells or systems containing thousands of millions of cells, are made up of functional parts: nuclei, mitochondria, plastids, etc. in the first case, and roots, leaves, limbs, hearts, lungs, etc. in the second.

6. All these functional parts can in some sense be regarded as parts of a mechanical whole and the question 'What is this part for?' is a reasonable one. They have a purposive character, using the word 'purposive' in its widest sense to include 'directive'.

7. It is shown that, as in the bacteriophage, the purposive character of the cell structures becomes equivocal at the molecular level.

8. One of the most important features of the cellular machinery, both in its structure and in the chemical events which take place in it, is its universality, e.g. carbon compounds, DNA and adenyl-phosphoric acid.

9. There is danger in looking for the origin of special features of organisms at too low a level. In DNA, as in computing machines,

and as presumably in brains, emergent properties are to be attributed to the configuration at a relatively gross level of complexity.

10. Whatever the theoretical basis of it, the concept of emergence is of value, so long as we treat as emergent phenomena associated with organisation which are not evident in the separate parts.

11. Systems with the complex emergent properties of organisms are few amongst the whole set of possible systems. Under the stress of natural selection, forcing species to traverse certain routes through the complex array of possible systems, species can evolve along a path of systems, always functional but passing during evolution to related systems with new functions. For this action of natural selection, heritable variation is a *sine qua non*.

4

The classification of objects and phenomena

THE PHENOMENA which we experience in everyday life seem to comprise material objects of different kind and different size, and events which involve these objects during the passage of time. Both the objects and the events seem to fall into recognisable kinds or classes, and one way of defining the objective of the natural sciences is to say that it is to determine the principles underlying the classification of objects and events.

Whether we have direct or indirect acquaintance with them, our world seems to consist of a multitude of objects. The individuality of these objects is not the result of an arbitrary mental division of our sensory field. It is not like the individuality we impose by defining, say, the interval between the ninth and tenth centimetres of a straight edge as an object for discussion. It is like the individuality of the projections and re-entrants in a curving line. In his *History of the Inductive Sciences* William Whewell[1] said of this matter:

What are these conditions which regulate our apprehension of an object as one? ...the primary and fundamental condition is that we must be able to make intelligible assertions respecting the objects and to entertain that belief of which assertions are the exposition. A tree *grows*, *sheds* its leaves in autumn, and *buds* again in the spring, *waves* in the wind, or *falls* before the storm. And to the tree belong all those parts which must be included in order that such declarations, and the thoughts which they convey, shall have a coherent and permanent meaning. Those are *its* branches which wave and fall with *its* trunk; those are *its* leaves which grow on *its* branches. The permanent connexions which we observe—permanent, among unconnected changes which affect the surrounding appearances,—are what we bind together as belonging to one object. This permanence is the condition of our conceiving the object *as one*. The connected changes may always be described by means of assertions; and the connexion is seen in the identity of the subject of successive predications; in the possibility of applying many verbs to one substantive. We may therefore express the condition of the unity of the object to be this: that assertions concerning the object shall be possible: or rather we should say, that the acts of belief which such assertions enunciate shall be possible.

About a particular object an indefinitely large number of particular assertions can be made and about it alone. But many assertions about one object can also be made about certain others. Objects are to be classed together into kinds. We may frame logical systems for the classification of them. But classification is also inherent in perception and is not of necessity always arrived at deductively or verbally. Indeed animals other than man behave as though they recognised classes. The behaviour of *Octopus* is consistent with recognition of the class of crabs. There is no reason to suppose that it is troubled about logic, definitions, and the meaning of words.

We do not classify by framing definitions. We perceive that objects are like certain others. When we examine such a class we can usually enumerate characters common to it. This does not mean that we first perceive the characters and then operate with them logically to define the class. It is in fact that we recognise classes or kinds. Any explanation of how this is done has only to meet this specification. We must be prepared to find that classification is carried out by us in several quite distinct ways at the same time.

Recognition of an individual and recognition of the class to which it belongs are part of the same process. Objects fall into classes because of the numerous characters they share in common and do not share with other objects. Even in the physical world, the relations of these classes are highly complex. From what we have seen, the classes of living creatures with their elaborate and often seemingly purposive organisation might be expected to show still more complex relationships. It is thus somewhat surprising that classification has made most progress in biology.

The classification of objects and phenomena

A classification may serve two different though related purposes:

(1) It may enable us to identify the class to which an object belongs.

(2) It may enable us to display the relationships of different kinds of object, their resemblances and their differences.

If we try to systematise these two processes of classification we obtain somewhat different results. For identification, we produce keys for objects, as in the taxonomic keys used for the identification of plants, or as in the group analysis of the chemical elements. The pattern of relationship between the objects grouped in such keys is scarcely made evident by them. Alternatively, for the display of relationships we use classification of a different sort, as in the periodic table of the elements and the so-called 'natural' classification of animals and plants. In the organic world, classifications of the first sort, the key, are associated with the name of Linnaeus; classifications which display natural relationship are associated with the name of John Ray.

The relation of these two kinds of classification is seen in the following examples given by a great taxonomist, the late W. T. Calman (1919):[2]

A. *Key to the species of part of the genus of barnacles Megalasma*

A. Carina projecting well below scutum	
a. Basal margin of carina as long as that of the scutum	*M. gracilis* (Hoek)
b. Basal margin of carina shorter than that of the scutum	
a. Carina transversely expanded at base	*M. gigas* (Annandale)
b. Carina not transversely expanded at base	
α. Sides of carina widened in lower third	*M. annandalei* Pilsbry
β. Sides of carina widened throughout its length	*M. pilsbryi* Calman
B. Carina not projecting below scutum	
a. Occludent margin of scutum nearly straight	*M. rectum* Pilsbry

b. Occludent margin of scutum convex	
a. Basal width of capitulum little less than one-third of its length	
α. Basal margin of carina shorter than that of scutum	*M. subcarinatum* Pilsbry
β. Basal margin of carina as long as that of scutum	*M. orientale* Calman
b. Basal width of capitulum no more than one-fourth of its length	
α. Numerous filamentary processes on dorsal surface of prosoma	*M. carinatum* (Hoek)
β. A pair of uncinate processes on dorsal surface of prosoma	*M. hamatum* Calman

B. *Classification of British Insectivores*

Phylum CHORDATA
Subphylum VERTEBRATA
Class MAMMALIA
Subclass EUTHERIA
Order INSECTIVORA

Family *Talpidae*	
Talpa europaea L.	Common Mole
Family *Soricidae*	
Sorex araneus castaneus Jenyns	Common Shrew
Sorex granti Barret-Hamilton and Hinton	Islay Shrew
Sorex minutus L.	Pigmy Shrew
Neomys fodiens bicolor (Shaw)	Water Shrew
Crocidura cassiteridum Hinton	Scilly Shrew
Family *Erinaceidae*	
Erinaceus europaeus L.	Hedgehog

A key is essentially an artificial, generally dichotomous, classification, successively dividing species into those with and those without some chosen character. Often a concession is made to the complexity of the features by which we recognise species by using sets of characters rather than individual characters. In using the key for identification of an animal or plant we thus follow a series of simple logical deductions:

All specimens with the set of characters *X* are *A*'s.
This specimen is an *X*.
Therefore it is an *A*.

We then proceed to the next set of characters; and our train of syllogisms finally arrives at some such statement as:

This is *Erinaceus europaeus* L., the common hedgehog.

The characters upon which the successive classes of a key are based may be arbitrary. Different keys for the same groups of species can be constructed on quite different characters. The features must of course be included in the total set which characterises the group, but the dichotomous classification may be as far from what would satisfy a naturalist demanding a natural classification as the table of group analysis of the metallic radicals differs from the periodic table. Undoubtedly the prime utility of this sort of key is that by its aid we can tell precisely what is the species to which our specimen belongs. In choosing the characters by which we differentiate each subdivision we are guided not only by characters of the specimen but by those which are easily seen and about which presence or absence can be stated with certainty; that is, the key conforms to the requirements of our mental machinery. The preliminary divisions of the key simply enable us to track the specimen down towards its species. Unlike individuals and species, we cannot necessarily make a very large number of independent assertions about each higher division. If there is in fact a 'natural' classification and the key is so chosen that it follows its divisions, we could of course make such assertions about all the species collected in the subdivision. But we are not compelled to make the key do this, though there is value in doing so. That is because systematic biology has not reached finality and the attempt to incorporate a new species into an existing arbitrary key may prove impossible; but this is less likely to happen the more closely the key conforms to a natural classification.

At first sight, artificial classifications by key may seem prosaically practical—like a telephone directory—and of little importance in the intellectual development of a science. That this is not so is best seen by considering the history of our classification of living things. Our modern ideas of classification arise from the work of John Ray in the seventeenth century. He took the older philosophical terms of genus and species and gave them the special taxonomic meaning they have for biologists today. As Raven[3] says, before this no attempt had been made by any student to define what constituted a species. Ray's paper[4] on *The Specific Differences of Plants* of 1674 begins:

> Having observed that most herbarists, mistaking many accidents for notes of specific distinction, which indeed are not, have unnecessarily multiplied beings, contrary to that well-known philosophic precept; I think it may be not unuseful, in order to the determining of the number of species more certainly and agreeably to nature, to enumerate such accidents and then give my reasons why I judge them not sufficient to infer a specific difference.

He showed how certain features of plants were the accident of 'the soil, the season, the climate or some other external circumstance', whilst their classification was to be decided by certain essential points of flowers and seeds, calyces and seed vessels. No real progress in any kind of classification could be made till species could be adequately described and it was Ray who first showed how this was to be done; on it rests the whole of our interpretation of the relationships of living creatures.

The importance of Linnaeus is different. Today he is often considered the author of binomial nomenclature. But in some form or other that is very old: barn owl, little owl, tawny owl are binomials. Linnaeus' binomial nomenclature was in truth a by-product. His great achievement was the linking of these names with no less than 10,000 descriptions of plants and animals, and carefully drafted definitions given by his Latin polynomial descriptive and diagnostic names; such was his *Potamogeton foliis oblong-ovatis petiolatis natantibus*: such was for him the true specific name. The task was colossal—and doubtless even he would never have attempted it had he not had the advantage of working originally with the impoverished flora and fauna of Sweden. It is one of many examples of how much we owe to the man of science who by accident or design attacks a problem in a very much simplified form.

Linnaeus stated his object in describing and naming things in these words: 'A rustic knows plants and so maybe does a brute beast, but neither can make anyone else the wiser. The botanist is distinguished from the layman in that he can give a name which fits one particular plant and not another, and which can be distinguished by anyone all the world over.' His binomial nomenclature arose because he was unconsciously engaged upon two divergent tasks. He was giving diagnostic descriptions, which, however condensed, necessarily tend to be long, but he also needed short designatory names which could be easily memorised and used in speech.

Linnaeus was concerned essentially with identification, and he used his classification as a key. For animals he based his classification on that of Ray, who was aiming at a natural classification. But in plants he constructed a purely artificial system based on enumeration of sexual organs. Though an artificial key may seem an arid thing, Linnaeus' system caught the eighteenth-century enthusiasm for the natural world, from the *Botanical Letters* of Rousseau to the time when it fired the young schoolmaster Alfred Russel Wallace to become a naturalist. For through easy and accurate identification we are brought directly into contact with natural objects, just as prosaic and detailed study may be the only way to enable a connoisseur to appreciate the work of an artist.

The purpose of a classificatory key is ostensibly to find the identity of a particular unknown individual. A natural classification differs from this in many particulars. In the first place it starts with the supposition that we know all about the things to be classified, and its object is to display the relations of these. We are guided to it first by an aesthetic impression of similarity. The higher orders of the classification do not differ from the lowest ones in this respect. A taxonomist may say at once about a new specimen, 'Whatever it is, I feel sure it is a member of the family Metrididae', even if he has great difficulty in saying why that is so. In fact statements about the higher classes are statements about any individual referred to: 'That is a mollusc', 'That is a snail', 'That is *Helix pomatia*', 'That is the individual from my vivarium'.

The classes and grades aesthetically perceived can be given precision by analysing our impression and extracting characteristics from it. By these we can define a mollusc, a gastropod, the genus *Helix*, and so on. Such definitions constitute a sort of template of 'all or nothing' features which we apply to our continuous aesthetic impressions of relationship. It is necessarily an approximation and historically requires revision from time to time to maintain its correspondence with our impressions. Moreover, the extracted characters of such a template are very far from comprising the whole of the evidence on which our aesthetic conviction is based. On the other hand such an attempt to extract defined characters, either for a key or to fit a natural classification, enormously enriches our aesthetic impression of the object.

As the taxonomy of plants and animals developed during the late eighteenth and early nineteenth century, it was perceived that there appeared to be a natural classification of a linear branching character to which successive subordinate divisions could be easily applied. After 1859 it was realised that a natural classification of this sort must inevitably be expected following Darwin's theory of evolution by natural selection. In fact, the manner of the origin of species by descent from common ancestors imposes a branching linear classification of the very kind which is most easily amenable to our linear logical arguments. It is this most happy accident which has enabled the biologist to proceed so easily towards classifying the most complex of all natural objects, living organisms. It is also responsible for his calling his classification 'natural'; in part this means that it is natural to his mind.

But it is the office of a natural classification to display all the relations of individuals, species and higher classes which it represents. The natural classifications of zoology and botany certainly assign a position to every known organism, but they do not display all their relations.

This deficiency is clear if we compare the classification of molecules by the organic chemist with the so-called natural classification of animals. The chemist can specify his different species of molecule with much greater precision than can the biologist his organisms. His structural formulae and his extended names are diagnostically admirable. But he has never succeeded, as Linnaeus did, in giving us clear, brief names for his species. There are no clearly defined genera, and clearly defined higher classes. Moreover, for convenience he is forced to coin terms. From guanidine to caffeine, to luminal and streptomycin, we enter a chaotic world of nomenclature without genera, in which each specialist gives a name at his convenience; often different specialists coining different names for the same substance. Worse still, in these intensely specialist days a research group will promote further contractions, DDT and DNA, a convenient but uncontrolled shorthand. Certain curious psychological principles seem to operate in the invention of these names; and these are quickly seized upon and used commercially for the construction of pseudo-scientific names for pharmaceutical and other products.

Linnaeus has at least spared the biologist that. But the fault is not with the chemist. His species certainly indicate relationships of higher and higher order. But any classification must essentially be multidimensional—just as it is in a set of dominoes. It cannot have the linear-dichotomous form applicable to the higher living things.

The contrast is easily seen if we compare two classes. In biology the molluscs and the insects comprise classes of distinct types. Individuals in each are rich in distinctive qualities, collections of qualities which have about them that peculiar character of uniqueness and improbability which tempts us to ask, 'Why should a mollusc be like that?'. The same thing can be said in organic chemistry about proteins, with their chains of amino acids, and also about the carotenoids, with their chains of isoprene links. Now we cannot find, and it would be impossible to conceive of, an individual which combined all the distinctive qualities of a mollusc and an insect. On the other hand, carotenoid–protein compounds, like the blue colour of lobsters, provide a common and important group of pigments.

We place objects in classes when they have the following property—that there is a large class of assertions about any of them such that any other object is very likely to have them almost all or almost not at all. Things are likely either to have almost all the characters we can find in a rabbit or relatively few or none of them. Intermediates in any direction are rare. In this way, rabbits and other organisms fall naturally into species. But we find that in turn certain species can generally be clearly grouped together by the common possession of a collection of characters absent in all other species. For among the exceedingly numerous characters that bind two individuals into a species there is a fairly large number of characters which they share with individuals of certain other species, and these related species constitute a unit of a higher order, the genus. A whole set of assertions can be made about all the species comprised in a genus which cannot be made about other species, and so on for higher groups: families, orders, and so on.

In the 'natural' classification of organisms, these successive classes are arranged in strict subordination so that no species is included in two different genera or orders; and the whole system branches from a single stem.

But it is quite easy in the physical world to find objects which can be classed, yet which do not exhibit this branching subordination of classification. The elements of the periodic table similarly fall into recognisable classes which cannot, however, be arranged in linear subordination; their relations require more dimensions for their expression. The same is true at a gross level in the classes of geographical features.

A classification convenient to the logical machinery of our minds requires two features. It must be a linear system which can be subjected to logical dichotomies, and the objects to be classified must be capable of precise definition. There is no arbitrary reason why the external world of which we are aware should condescend to meet these requirements. Nevertheless, to some degree it does so. At the atomic level the indistinguishable character within their classes of individual particles, atoms and molecules allows us all the precision of definition demanded by the logical machinery of our minds. But the classes do not allow linear subordination. On the other hand our so-called natural classification of living things shows linear subordination of the classes. Does it allow us to make precise definitions?

In natural history, both keys and natural kinds of classification presume the reality of species. Are there really such things as species of animals and plants? The word 'species' is only a substitute for the word 'kind' which has been made more exact by very careful observation and the application of an agreed set of rules and description. Whenever by reason of their occupation, particularly with sport or agriculture, men have been led to make a minute and shrewd examination of organisms, they soon distinguish 'kinds' with some precision.

Though there are undoubtedly some very difficult cases where species are hard to define or grade into each other, in the very great majority they are quite clear, and when one biologist speaks to another about species he is not speaking about something which does not exist. If there was no such thing as species the natural world would present quite a different appearance to us from what in fact it does.

To some extent this is also true of the genus and higher orders of classification. It is sometimes held that, compared with the species, these are mere abstractions. But it is significant that the common man identifies creatures by genera before experience teaches him to identify species: 'There is a thrush' precedes 'There is a missel-thrush'; and Linnaeus himself was originally concerned with self-evident genera—to which he then applied specific differentiae. That is why we talk of *Amoeba proteus* and *Homo sapiens*.

But there are in fact two questions, of which only the first is: Do species exist? We can also ask: Does their existence require that we must be able to make a precise definition of them? Just over a hundred years ago Whewell gave a most valuable discussion on this question. It was the more so because it was made before, though not so very long before, the publication of *The*

Origin of Species, at a time when species themselves were being critically recorded and classified, yet before the complication of evolutionary interpretations masquerading as premisses. Whewell[5] points out that in fact we cannot define species, genera or other groups which we take to be natural, by precise definitions. We assign organisms to species, species to genera and so on by the general weight of all their characters. Individual characters may be gainsaid in individuals which we nevertheless unhesitatingly assign to the class. Whewell says:

These views—of classes determined by characters which cannot be expressed in words—of propositions which state, not what happens in all cases, but only usually—or particulars which are included in a class though they transgress the definition of it, may very probably surprise the reader. They are so contrary to many of the received opinions respecting the use of definitions and the nature of scientific propositions, that they will probably appear to many persons highly illogical and unphilosophical. But a disposition to such a judgement arises in a great measure from this—that the mathematical and mathematico-physical sciences have, in a great degree, determined men's views of the general nature and form of scientific truth; while Natural History has not yet had time or opportunity to exert its due influence upon the current habits of philosophising. The apparent indefiniteness and inconsistency of the classifications and definitions of Natural History belong, in a far higher degree, to all other except mathematical speculations: and the modes in which approximations to exact distinctions and general truths have been made in Natural History, may be worthy of our attention even for the light they throw upon the best modes of pursuing truth of all kinds.

But...though in a Natural Group of objects a definition can no longer be of any use as a regulative principle, classes are not, therefore, left quite loose, without any certain standard or guide. The class is steadily fixed, though not precisely limited; it is given, though not circumscribed; it is determined, not by a boundary line without, but by a central point within; not by what it strictly excludes, but by what it eminently includes; by an example, not by a precept; in short, instead of a Definition we have a Type for our Director.

A Type is an example of any class, for instance, a species of a genus, which is considered as eminently possessing the characters of the class. All the species which have a greater affinity with this Type-species than with any other form the genus and are ranged about it, deviating from it in various directions and different degrees. Thus a genus may consist of several species which approach very near the Type, and of which the claim to a place with it is obvious; while there may be other species which straggle further from this central knot, and which yet are clearly more connected with it than with any other, and even if there should be some species of which the place is dubious, and which appear to be equally bound to two generic Types, it is easily seen that this would not destroy the reality of the generic groups, any more than the scattered trees of the intervening plain prevent our speaking intelligibly of the distinct forests of two separate hills.

As a naturalist I feel that what Whewell says is both correct and illuminating. But we must guard against confusion about what is the nature of the type. The type we use in recognition as our director is not a material object. It is the machinery of our trained perception. It is that modification of our perceptive faculty which has been produced by the integration of all our past experience connected with objects of the class. We may for convenient illustration of it choose a material object, a type specimen, which corresponds closely with our perceptual type. But our recognition of the species was prior to selection of the example.

The importance of this perceptual type is very great. What has been said about it applies not only to organisms, but to every kind of physical object and every kind of event which can be assigned to any kind of class.

Our notion of a species is thus bound up with the way we recognise it. It is really a statistical concept—or, as I prefer to call it, a book-making notion; for the recognition of species has more to do with backing a fancy than with the actual working out of correlation coefficients. As with book-makers and art critics, the judgement of a taxonomist depends upon the training experience has given to his perception and insight. That is why Tate Regan[6] did not beg the question when he defined a species as an assemblage of organisms recognised as a species by a competent taxonomist.

Our 'book-making' definition of species implies that, though they certainly exist, there may be odd individuals which are difficult to place within or without the species. It does not even exclude the possibility of instances in which the odd individuals are so numerous and varied as to make species unrecognisable. It is the fact that species are described by type and not by definition that undermines the fallacy that the existence of intermediates which are difficult to classify renders a whole classification meaningless.

The whole notion of species arises from our everyday observations of animals and plants. Definition of scientific phenomena should be based on the phenomena as we see them. We have no business to base our definition on ideas of what we think phenomena *ought* to be like. The quest for such touchstones seems to arise from a private conviction that simple laws and absolute distinctions necessarily underlie any connected set of phenomena, just as Kepler thought that the number and distances of the planets from the sun could be related to the five regular geometrical solids. Such convictions are often useful by suggesting what observations or experiments we might make next. They may or may not prove to approximate to the observed phenomena. But they are certainly not necessarily justified. Simple relationships and well-defined classes are indeed to be found in the physical world. The classes of crystals, the relation of the electron shells to the position of atoms in the periodic table, and the existence of three states of matter are systems of simple relationship which are certainly not imposed on the natural world by the human mind. But the desire to impose simplicity upon our experience is suspect. Whilst it may sometimes be that they do give a fair approximation to an underlying natural plan, the reason we are indulgent towards the supposed existence of simple relations and absolute distinctions in natural phenomena is often because of their great convenience for our own logical processes: we may be attributing to the natural world a characteristic derived from our own mental machinery. As Whewell[7] says:

We entertain a conviction that there must be, among things so classed and named, a possibility of defining each.

Now what is the foundation of this postulate? What is the ground of this assumption, that there must exist a definition which we have never seen, and which perhaps no one has seen in a satisfactory form? The knowledge of this definition is by no means necessary to our using the word with propriety; for anyone can make true assertions about dogs, but who can define a dog? And yet if the definition be not necessary to enable us to use the word, why is it necessary at all? I allow that we possess an indestructible conviction that there must be such a character of each kind as will supply a definition; but I ask on what this conviction rests.

I reply, that our persuasion that there must needs be characteristic marks by which things can be defined in words is founded on the assumption of *the necessary possibility of reasoning.*

Whatever their nature, it is the species of organisms which we arrange in one or another sort of classification. Both a key and a 'natural classification' display all the species; and these are in each case subordinated under groups of higher order, and these again of higher order, and so on.

This special feature of the higher organism depends on the fact that under natural selection each species has followed a highly individual path of evolution. The probability that two individuals should have arrived at the possession of all the special features of a mollusc by totally different evolutionary routes seems indefinitely low.

The defect of the so-called 'natural classifications' is that, though they should display *all* the relations between the different kinds of organism, they fail to do so. Indeed their failure provides one of the greatest difficulties in the organisation of the training of young biologists. In immediate post-Darwinian times, the matter seemed easy. In zoology the training consisted of courses of lectures on evolutionary morphology. All the species were assigned a place and their taxonomic features discussed. But today, courses on behaviour, functional morphology, physiology, the cell and ultrastructure, biochemistry, genetics and ecology discuss highly important relationships between organisms which seem to cut right across those of the old taxonomic morphology.

Part of the difficulty is that today we are concerned with many more levels of morphology, in some of which the rules of gross morphology do not apply. The different kinds of likeness visible when we compare the features of living organisms may be said to fall into three classes: homology, analogy and chemical identity. The first two of these were given precision in the middle of the last century by Richard Owen.[8] He defined them: a homologous organ is the same organ in different animals under every variety of form and function; an analogous organ is a part or organ of an animal which has the same function as a part or organ of a different animal. Chemical identity is a property that has come to light during the present century through the rise of biochemistry. Particular sorts of complex molecule recur in organisms which seem to show no evolutionary relationship. Pharmacologically, leeches and men happen to show an astonishing similarity of chemical systems. But leeches are no evolutionary relatives of ours. In the same way the fact that man and the unicellular creature *Paramoecium* both have the complex and physiologically important molecule acetylcholine does not argue for a common ancestry in the same way that the presence of a hand does for a man and for a mole. There is good reason to suppose that, for all their apparent complexity, the recurrence of such molecules in different organisms is a reflection of the limited number of possible kinds of molecule available to organisms for their construction. That arises directly from the limited number of elements with their unique properties which are available in the periodic table.

At the chemical level of specific characters that same multi-dimensional series of relationships emerges which is so obvious in the physical world. Consequently the 'natural classification' runs into difficulties. That is very evident in the classification of bacteria. Here specific differences often rest on the presence or absence of some chemical process which is in fact the sign of the presence or absence of some enzyme.

Here are some generic characters given in a recent text-book of bacteriology:[9]

Genus *Nitrosomonas*: Cells ellipsoidal, non-motile or with a single polar flagellum, occur singly, in pairs, short chains, or irregular masses which are not enclosed in a common membrane. Oxidise ammonia to nitrite more rapidly than the other genera of the Tribe.

The taxonomic weakness of the structural characters of bacteria has long been recognised; and it would be rash to suppose that even the whole set of diagnostic characters given above could not have been evolved together on more than one occasion. Nowhere do we have taxonomic uniqueness comparable with that of the pentadactyl limb. All the same, the difference between these elementary structural features and those which influence the construction of a 'natural' classificatory system is to some extent a matter of degree.

But the most important systems of relations which the 'natural classification' leaves undisplayed are those provided by analogy. After 1859 analogical resemblances, the mere functional resemblance of our hand to the claw of a lobster, were brushed aside. But within the animal kingdom functionally analogous organs may achieve a remarkable similarity in quite unrelated creatures.

The analogy is closest where the imposed functional specification is the most detailed. That is the case with those organs that receive the unsorted information from the environment. Eyes must sort visual patterns from the field of incident light. Likewise, the sorted information must be fed into predictor machines of a nervous system, again with a detailed specification, such as we require for the construction of calculating machines. In contrast, the imposed specification for movement in animals is much less detailed. Animals achieve movement in all sorts of ways. In fact the characteristic 'look' of an animal is largely due to its motor machinery. The eyes and the brains make for similarity. An octopus is obviously staring at you—it is its arms that make it so inhuman and uncouth.

Functional analogy is far from trivial. We do not hesitate to interpret the action of our nerves in the light of experiments done on the giant nerve fibres of squids, or to compare the structural basis of the process of learning in *Octopus* with that of a mammal.

I once discussed this question elsewhere[10] and arrived at the following conclusion. The organism is built up of standard parts with unique properties. The older conceptions of evolutionary morphology stressed the graded adaptation of which the organism is capable, just as putty can be moulded to any desired shape. But the matters we have discussed lead us rather to consider the organism as more like a model made from a child's engineering constructional set: a set consisting of standard parts with unique properties, of strips, plates and wheels, which can be utilised for various objectives such as cranes and locomotives. The existence of such sets in the elementary particles, atoms and molecules of the physical scientist is the basis of the kind of likeness I have called 'chemical identity'. Models made from such a set can in certain respects show graded adaptability, when the form of the model depends on a statistically large number of parts. But they also show certain severe limitations dependent on the restricted properties of the standard parts of the set. Moreover, in this universe of ours any functional problem must be met by one or other of a few possible kinds of solution. If we want a bridge, it must be a suspension bridge, or a cantilever bridge, and so on. And the engineer who constructs the bridge must choose whichever of these solutions he can best employ with the standard parts at his disposal. These limitations are the basis of analogy. In the design of a bridge there are in fact three elements: the classes possible in this universe; the unique properties of the materials available for its construction; and the engineer only takes third place by selecting the class of solution and by utilising the properties of his materials to achieve the job in hand. He is in a sense merely executing one of a set of blueprints already in abstract existence, though it requires insight to see that the blueprint is there. We can apply these ideas to the construction of the living organism. Like all material structures they must conform to certain constructional principles. The standard parts available for the construction of organisms are the units of matter and energy which can only exist in certain possible configurations. Like the engineer, natural selection takes third place by giving reality to one or other of a series of possible structural solutions with the materials available.

To display all the relationships to be perceived in the natural world we would need a multidimensional system including representations of all possible material objects and configurations. The evolution of living creatures can be represented in such a system as a linear-branching sequence of objects or organisms. The character of these at each point in the sequence is determined by the operation of natural selection. Considering these organisms simply as material configurations, those of common origin will show similarities of plan, the homologies of Richard Owen which are seen in the higher organisms. They will also show structural similarities due to what I have called chemical identity: the repeated and independent recurrence of chemical units, and structures built from them. Molecular similarity may be reflected in structural features of a higher order. These structural features of organisms are the independent natural consequence of their construction from actomyosin fibrils, in the same way as snow crystals repeatedly generate complex patterns of the same form. Such resemblances may cut right across the evolutionary series of configurations.

But this by no means exhausts the relationships to be perceived in the natural world. Our multidimensional system must include not only configurations but also the events in which they participate: it must include what happens to configurations and what particular configurations do. If we take the event which is a river flowing during a certain time we can note that this dynamic system is one about which numerous assertions can be made. We can note, for instance, that it is part of a steady state completed by the carriage of water from the sea to the atmosphere, to the rain, to the river. It is here that we find the classes to which scientific research endeavours to assign particular phenomena; that class to which a phenomenon belongs. The event of a liquid flowing on a solid surface shows certain common features at all scales of size and with very different materials. Relaxation oscillations embrace events in classes of objects which are as widely separate as a flag flapping in the wind, a beating heart, and an automatic cistern. It is the existence of these classes which renders possible the construction of mechanical models: and it is the images of these classes in the mind which enable us to apprehend intuitively through analogy the nature of a system we investigate.

I have noted that within the multidimensional system which must represent the possible relations of the form of objects through the operation of natural selection there is a linear branching system of particular objects corresponding to the course of evolution of living organisms. A particular linear system is also to be found in the system of events. To display all the relations of events we need a multidimensional system. Some of the possible objects concerned in this system have emergent properties, such as the power of replication of nucleic acids, and the power of sorting information in calculating machines or central nervous systems. The events in such systems seem directed to a future end. In living organisms they can be classified according to function by the physiologist.

This functional classification cuts right across the so-called 'natural classification' of organisms by structure. Nevertheless, as a classification it is clear: we can make numerous common assertions about vision in vertebrates, in *Octopus* and many other animals. We can also make many similar assertions about photography.

If we look at the chapter headings of a text-book of human or comparative physiology, we can find functional classes. The organism is a whole and the primary function is to get material to maintain and get more of it. To that end there are organs for movement, limbs and muscles; sense organs for collecting information; nervous predictor and controlling machinery; machinery for nutrition and for respiration; cellular metabolic machinery; homoeostatic machinery for maintaining constancy of conditioning in the body; reproductive machinery. All these functions provide specified requirements which the organism can only meet by the limited number of possible engineering solutions.

Because all these systems are necessarily integrated in the whole organism, this functional classification shows linear subordination of classes. The organism moves by limbs, these are operated by muscles, these have a contractile machinery of actomyosin fibrils, and so on. It is true that in the simplest organisms we have organs of multiple function. The muscular wall of a sea-anemone subserves both digestive movements and locomotion. But the functional classification concerns a functional aspect of each organ rather than the whole organ itself, and it is that which stands in a linear subordinate relationship, though in the 'higher' animals individual organs tend to become identified with one particular function.

Though linear subordination of classes has quite a different character in the so-called 'natural' and in the functional classification, they both may be presumed to owe this character to the operation of natural selection: in the one case through a common ancestor, in the other through the enforced integration of the organism which must be maintained for survival. In each case, moreover, the actual examples—the surviving species and the particular functioning of a particular sort of individual—represent a pattern of realised possible states during the course of evolution, in a universal set of possibilities.

The physiologist is concerned with the analysis of events. He has more than one method open to him. He may analyse function, that is, determine the 'purpose' of some function in the economy of the whole animal. But he may also study the relation of an event to its associated structures, just as Keith Lucas[11] and Adrian[12] studied the class to which the nervous impulse belonged. Such problems comprise the 'general physiology' of W. M. Bayliss[13] as opposed to 'human physiology', and 'comparative physiology' which are particularly concerned with the ascription of function.

But of these two approaches that of 'general physiology' is the more important. In any science it is by the application of the classified features of events to the classified configurations of objects that models are made, at least in the mind; and the production of such models is an essential part of the scientific method. Without such models and their analogies we have no moorings, no plan and no common sense.

SUMMARY

1. Objects are recognised as individuals and also as members of a class or classes. These two types of recognition are in fact an outcome of the same process. Classification is inherent in perception and indeed animals behave as if they recognise classes.

2. It is somewhat surprising, since living creatures have such

elaborate and seemingly purposive organisation, that classification is the aspect of biology that has so far made most progress. A biological classification serves two different though related purposes: (*a*) to enable us to identify the class to which an object belongs; (*b*) to enable us to display the relationship of different kinds of object (a 'natural' classification).

3. Modern ideas of classification arose from the work of John Ray, who first showed how species could be accurately described. Till this was done no real progress in any kind of classification could be achieved. Linnaeus was essentially concerned with identification and he used his classification as a key. His binomial nomenclature arose because he was unconsciously engaged upon two divergent tasks: (*a*) that of providing diagnostic descriptions and (*b*) that of providing short designatory names.

4. With the publication of *The Origin of Species* in 1859 'natural' classifications acquired a new significance. For it soon came to be realised that a natural classification of a linear branching character was inevitably to be expected from Darwin's theory of evolution by natural selection. The manner of the origin of species by descent from common ancestors imposes a branching linear classification of the very kind which is most easily amenable to our linear logical arguments.

5. But our natural classifications are only partly successful in that they display only some, not *all*, of the relations of individuals, species and higher classes. They do, however, assign a position to every known organism. By comparison, in his classification of molecules the organic chemist can specify his different species of molecules with far greater precision than the biologist his organisms. This he does by his structural formulae and his extended names, which are diagnostically admirable. But he has never succeeded in giving clear brief names for his 'species'. This is because there are no clearly defined genera and higher classes, for the objects which the organic chemist has to classify have not been produced by the process of natural selection. So in organic chemistry we enter a chaotic world of nomenclature without genera, and 'nicknames' without coordination or consultation.

6. To some extent the external world meets the logical machinery of our minds. Yet the problem of making precise definitions, posed by Whewell, remains. Both genera and species are 'real' in the sense that we have a concept of 'type' which can act as our director and which is the machinery of our trained perception. The notion of species is bound up with the way we recognise it and is really a statistical or book-making concept. As with book-makers and art critics, the judgment of a taxonomist depends on the training experience has given to his perception and insight.

7. Although the external world is in some sense and degree amenable to the logical processes of our minds, we must beware of assuming that simple relationships, which are so much easier for our minds and theories to cope with, are necessarily more likely to correspond with reality. The desire to impose simplicity upon our experience is suspect.

8. The kinds of likeness we find when we compare the features of living organisms fall into three clases: homology, analogy and chemical identity. While a natural classification can cope reasonably well with the first, it runs into difficulty with the others. At the chemical level specific characters show that same multi-dimensional series of relationships which is so obvious in the physical world. This is very evident in the classification of bacteria, where specific differences often rest on the presence or absence of some chemical process which is the sign of the presence or absence of some enzyme. Similarly with analogy. Analogical similarities tended to be brushed aside after 1859. Here again, as in the case of chemical identity, resemblances are often due to there being only relatively few possible ways of building a machine for a given purpose, and these ways are the expression of the fact that the organism is built of a relatively few standard physical parts with unique properties.

5

Methods in the unrestricted sciences

WHAT WE SEE, what we learn, and what we learn to see gives each of us a great image of the natural world. We apply to this image a system of hypotheses which we try to make self-consistent, and by which we purport to explain phenomena. Both our image and our system of hypotheses are contemporary products which change with time. Further, for each of us they are personal. They are detailed in our own field of interest, and in more distant fields they are not only less detailed but, like that from the celestial galaxies, the information received is apt to be increasingly out of date with distance.

Starting from some phenomenon, part of the scientific method as practised by every man of science is the framing of a new hypothesis which extends or modifies this system, followed by rigid deduction of consequences which can be verified by observation.

In making a new hypothesis he follows the principle of economy of hypothesis: he chooses as the most probable that which according to his lights he deems to be the simplest, to involve the fewest assumptions. He goes further than this. Faced with the welter of phenomena presented to his apprehension, he tends to choose for examination the simplest and most clear-cut. Experimental control of conditions is directed to simplifying phenomena to be investigated in this way. But search for the simplest phenomena for the scientific attack gives that attack a systematic bias in favour of the investigation of simple systems and the simplest aspects of complex ones. All this is to the good so long as we are constantly aware that complex phenomena may not only be complex but may include aspects of nature systematically excluded by our concentration upon simple phenomena.

It is here that the restricted sciences differ from the unrestricted. Because physics has excluded from its domain all but a selected few of the wealth of natural phenomena, its initial assumptions are few and clear-cut compared with those of other branches of knowledge. Consequently, in working out hypotheses, very long chains of deductive reasoning can be made, the results of which only occasionally require confirmation by observation.

The procedure in the restricted sciences can be illustrated by Newton's determination of the velocity of sound. Newton first developed his theory of the propagation of a pulse in an elastic fluid. He then resorted to experiment to test the predicted velocity. In fact the experiment failed to confirm the prediction because, as Laplace showed more than two hundred years later, his calculations treated the wave of compression as isothermal, whereas it is necessarily adiabatic, and Laplace's adiabatic correction led to tolerably complete agreement between prediction and subsequent experiment.

Newton's experiments were thus preceded by a mathematical model. The assumptions of that model were in turn derived from contemporary knowledge of the properties of matter.

Provision of an explanatory hypothesis involves the construction of a mental model. That model is built up from our contemporary apprehension of the properties of objects; that is, of the classes of events associated with particular configurations. These classes are idealised, and thereby given a precision which allows logical development of their consequences. In a sense, all these logical developments may be tautologies in that, as in a system of geometry, any one proposition implies the truth of the others. But, as in geometry, because of the poverty of our insight, we do not see these developments—except sometimes by a rare intuition—and to have them logically displayed in the mathematical model increases our knowledge. Thereby we can look for new and unexpected phenomena which greatly increase the specification of the class to which events belong.

Such models begin with the images of the properties of objects. The extent to which they are developed mathematically varies.

It may be great in the 'mechanical models' of physics after the fashion of Lord Kelvin's, or slight in the mud-and-water models that geographers use to analyse the progress of denudation. But models are rarely actually constructed.[1]

Because the images of the model are ultimately derived from concrete experience, there is danger that irrelevant properties of one system may be carried over to another. These difficulties would be avoided in a purely ideal model based simply on observed relationships such as, I believe, Einstein and mathematical physicists contrive to obtain. In such ideal models there is no danger of bringing into the hypothesis features which have as little to do with the phenomena to be described as the ivory of billiard balls has with the molecules of the kinetic theory.

When we turn from the restricted field of modern physics to the unrestricted study of phenomena, even in the physical world, the multiplicity of variables and the repeated obtrusion of new phenomena greatly limit the extent to which a mathematical model can be developed. Just because there are so many variables and unknown factors, the predictions of meteorology still lag far behind those of nuclear physics, for all its great importance. In all complex phenomena the number of initial assumptions for any hypothesis tends to be very high, particularly in biological hypotheses. Consequently their liability to error is great and, to borrow a physicist's phrase, their 'half-life' is short. Models of the structure of protoplasm and the lactic acid models of muscular contraction which were current in the 1920s make strange reading today. Particularly in biological advance, our hypotheses are perpetually breaking down.

I have said that we tend to choose the simplest phenomena for investigation, for it is of these that simple hypotheses can be constructed. But, as with other razors, one must take care not to cut oneself, even with William of Occam's. Our method tends to cut us off from the investigation of complex phenomena and from the examination of systems of a higher order. It is not certain that the remarkable emergent properties of such systems can ever be wholly deduced from the study of elementary systems alone, and we certainly cannot wait for this to be determined. If we wish to gain the maximum of knowledge we must study all systems of all grades of complexity all the time.

The principles of investigation applicable to simple systems need modification when applied to those of a higher order, particularly in biology, through the influence of natural selection upon them. If in a homogeneous chemical solution we find that some reaction takes place when it is exposed to light, and that the reaction is proportional to the amount of light, we can make a hypothesis of high probability that we are dealing with a simple photochemical reaction. During the early days of the analysis of animal behaviour, Jacques Loeb[2] claimed to show that the movements of certain animals are directly proportional to the quantity of light received. He argued from this that, as in a simple physical system, the movement of these animals must be directly controlled by a simple photochemical reaction. On such experiments as these he endeavoured to erect an entire deterministic system of animal behaviour.

It is important to remark that within the retina of the eye it is reasonably certain that photochemical reactions do, in fact, take place. That is not the point at issue. The question is whether simple proportionality of response of the animal to the quantity of light allowed Loeb to assume the existence of a simple photochemically controlled system as he would be justified in assuming for a test-tube experiment on a homogeneous solution.

Loeb was wrong. Experiments with flickering lights and a closer examination of the animal's movements showed quite clearly that the nervous machinery of such responses is not only complex but is in fact carefully adjusted to give a direct proportionality between stimulus and response under all normal conditions. The animal has, so to speak, gone to great trouble to produce a simple result.

The reason for this simplicity of relation is not far to seek. In searching for simple reactions for experimental analysis, Loeb found ready to hand those which happen to be concerned with the orientation of animals in space. But in such orientation reactions there must of necessity be a simple relation between stimulus and response. In our own bodies, unless the tension in our muscles precisely balances the weight on our limb, the body will either shoot upwards or collapse. That is, unless the nervous system is constructed and most carefully adjusted to give just such a simple relation, the machine will not function and cannot survive. Thus, in fact, natural selection enforces a specious simplicity on biological systems which may deceive us in constructing our hypotheses.

Such difficulties of the scientific method in biology arise because in living organisms our subjective estimate of what is probable becomes upset. In simple physical systems we are accustomed to a particular class of events depending inevitably upon the presence or absence of a particular class of object. Remove the object and no event of the class takes place. Replace it, and the event of the class can recur. But this does not rule out the possibility that an event of that class might under certain conditions be produced by objects of a different sort. There may be alternative means of meeting the specification of the event. Light of wavelengths 5890 and 5896 tenth-metres is commonly associated with the presence of sodium. It might, however, be produced by a monochromatic illuminator. But in the ordinary course of nature, for instance in the chromosphere of the sun, we deem the presence of such illuminators, set to give just these values, so unlikely that we assume at once that such light is due to the presence of sodium.

Such judgement can be rendered invalid by human intention or deceit. The astronomer, Sir William Herschel, was asked by the ladies of the court of George III to show them the planet Saturn. The night being cloudy, he set up a distant light masked by a card with a hole cut to the shape of Saturn. He said that the image through the telescope might have deceived even a skilful astronomer. He was no fraud and told the ladies what he had done. But in more recent times at Piltdown in Sussex a famous and deliberate fraud was made. Part of an ancient skull, the artificially stained and carefully broken jaw of an ape, a filed and shaped tooth, and various artifact objects and implements were, with appropriate stratigraphical legerdemain, made to appear the true relic of the long-promised Pliocene man, *Eoanthropus dawsoni*. The brilliant detection of this fraud depended upon the fact that our contemporary specification exacted of Pliocene human remains has become both more detailed and different from that expected in 1912. From being an ancestor, *Eoanthropus* became an anomaly. One of the salutary benefits of the history of science to the advancement of knowledge is the demonstration that what were once hidden axes to grind gradually become manifest as halberds.

But our natural expectation of likelihood is not only vulnerable to deliberate human falsification, in illustration and in fraud. It is also upset in living organisms. It is not merely that these, like machines, have that improbable air of design about them. In biological systems, owing to selection to meet a functional specification or need such as respiration, we constantly find alternative systems adapted to meet that specification simultaneously present in the same organism. Consequently, unlike our sodium example, if one of these systems is rendered inoperative the subsequent event may still retain the features which place it in the original functional class.

I have just discussed the influence of light on animals. Many simple worms avoid light and collect in darkness. These worms have eyes which normally control their reactions to light. But remove the eyes and the animals still collect in the dark. The skin of these worms is sensitive to light, and this takes charge when the eyes are absent. Again, the worms usually travel directly away from a source of light. But if the animal is put in a gradient of illumination so arranged that there is no directional quality in the light, the worms still all manage to collect in the dark through an ingenious trick of behaviour.[3] In the same animal there is no single mechanism for 'collecting in the dark'. There is a whole set of mechanisms. Natural selection has seen to it that if one fails another stands ready to take its place.

Although it is especially in biology that we meet such difficulties of analysis, they are not peculiar to biological systems.

They are also to be found in machines in which several functional systems are designed to meet the same specified end. The simple physical systems are simply limiting cases where the occurrence of alternatives strikes us as indefinitely improbable.

When analysing phenomena the biologist repeatedly sees his hypotheses collapse, and it is more difficult for him to exclude alternatives than it is for the physicist. Repeatedly he has to make new hypotheses about structure and these require fresh specification about function. That is very clearly seen in the history of our knowledge of nervous action and the function of the brain. In Whewell's time it was already known that something was transmitted by nerves. Cut a nerve, and the will has no power over the muscle. Cut a nerve and sensation is lost. It was also known since the experiments of Galvani late in the eighteenth century that nervous tissue was peculiarly susceptible to electrical stimulation. But in the 1820s Johannes Müller[4] had concluded

> that the vital actions of the nerves are not attended with the development of any galvanic currents which our instruments can detect; and the laws of action of the nervous principle are totally different from those of electricity.

On this Whewell[5] argues:

> That the powers by which the nerves are the instruments of sensation, and the muscles of motion, are vital endowments, incapable of being expressed or explained by any comparison with mechanical, chemical and electrical forces, is the result which we should expect to find, judging from the whole analogy of science; and which thus is confirmed by the history of physiology up to the present time.

Of course in the 1820s Johannes Müller stood no chance of detecting an action current of 10 millivolts nor had he the optical equipment to see the axons in which it ran. It was all outside the contemporary sense-spectrum. It will be seen that Whewell rejects the connection of nervous action with electrical currents; and then, with a suitably phrased emergency exit, boldly claims that nervous action is a vital endowment which cannot be expressed in physical terms.

But almost at the same time that Whewell was writing,

Du Bois-Reymond was removing the negative evidence that fortified his conclusion. Du Bois-Reymond showed that it was possible to measure the speed of nervous transmission and that the transmission of excitation was accompanied by a wave of electrical change. Gradually by further experiments the specification of the phenomenon of transmission became more detailed. At the turn of the century quantitative study of electrical excitation enabled Nernst[6] and subsequently A. V. Hill,[7] to consider a model of nerve consisting of a membrane across which there was a difference of electrical potential. By a mathematical model based on this, predictions about nervous excitation were made, which were verified by Keith Lucas. This same process of increased specification has continued, till today we make 'anatomical drawings' of the molecular organisation of the cell surface; and we know a very great deal about the electrochemical machinery of nerve cells.

The work of Keith Lucas and of Adrian[8] showed that the size and quality of a stimulus could not alter the size and quality of a nervous impulse. In making hypotheses about how the whole nervous system worked, the only way in which gradation of sensation or of muscular response could be envisaged was by varying the number of impulses and the number of channels employed. The individual impulse was 'all or nothing' and could not itself convey graded information.

All this study of conduction in nerve fibres was still a long way from the study of what went on in the brain. But immediately after the First World War attempts were made to see how far nervous impulses passing through networks of nerve fibres such as were to be seen in the brain could account for the phenomena of central nervous action. Of themselves they could not do this, and various subsidiary hypotheses about the properties of nerve junctions and of inhibitory action were called in to give partial help. One serious attempt on this problem was made in France by the physiologist Lapicque.[9]

The year 1918 and the following decade saw a rapid development of electronic technique. That enabled physiologists to observe on a scale never before possible electrical changes in the nervous system. It also endowed them with mental models derived from electronic circuits. One of the properties of these was duration of discharge and selective tuning to certain frequencies. The excitability of a nerve fibre was known to be maximal for an electric current of a particular duration. Lapicque boldly suggested that the controlling factor in selective nervous transmission, in brain and body, was due to what he called 'isochronism', a selective 'tuning' of the excitability of a muscle, or of one nerve cell, by another. An immense scientific literature accumulated on this; and its conclusions were in the end brilliantly and thoroughly shown to be worthless.

But at this very time a new discovery changed the whole accepted basis for central nervous models. It was found that nerve cells excited others and various tissues of the body, not directly by their electrical change but through the secretion of certain highly specific chemical substances. Since then we have come to know that nerve fibres very commonly act not merely as cables for electrical transmission but even as channels for the direct passage of specific chemical secretions. Further, new electrical studies have shown that the action current is by no means the only kind of electrical effect to be found in nerves. There are slow changes of potential, particularly at nerve junctions. There is some evidence of conducted disturbances other than the classical 'all or none' action current. Today the physiologist is faced with an embarrassing variety of phenomena[10] out of which to build his model of central nervous action: and he still knows that his notions may have to be changed by empirical observation. As Grey Walter recently remarked:[11]

> One of the most sobering, even humiliating, facts in the whole of brain physiology is that scarcely a single phenomenon discovered by study of electrical activity of the brain—the EEG—was foreseen or predicted by physiologists, and indeed few of these electrical effects are really understood today.

The history of our study of nervous action shows most of the special features of the scientific method in biology: the short half-life of hypotheses; the detailed and increasingly detailed specification of structure and function; the simultaneous alternative functions of the same structure; the implications of the alternating physiological and anatomical models; the fact that the primary goals of investigation concern configuration well above the atomic level.[12]

So far we have been chiefly concerned with methods by which a scientific attack upon a problem is developed. We now come to the primary question: how is it that we come to take up a particular problem? It will already be apparent that we start from our great accepted framework of hypotheses about the natural world. That framework is contemporary and is profoundly influenced by judgments of value, particularly the value of the reported work of others. Whether we like it or not, authority influences our judgement enormously.[13] A good illustration of this is the history of the determination of the chronology of the history of the earth and its rock formations.

The attack upon such a problem was deeply affected by contemporary tradition. Both authority and tradition have a historical background. This is not even the same in different sciences, a frequent cause of violent disagreement.

In the seventeenth century ancient chronology was based upon the assumption of the historical accuracy of the Book of Genesis. It is at first sight a strange thing that Isaac Newton[14] at the same time that he was engaged on his great work on celestial mechanics spent an enormous amount of labour on chronology on that assumption. He says:

> I have drawn up the following chronological table, so as to make chronology suit with the course of nature, with astronomy, with Sacred History, with Herodotus, the Father of History, and with itself; without the repugnancies complained of by Plutarch. I do not pretend to be exact to a year: there may be errors of five or ten years and sometimes twenty, and not much above.

Newton accepted the authority of the Old Testament absolutely. On this basis certain dates could be fixed almost exactly. He arrived at 1059 B.C. for the date that David was made King of Judah, which is only a small correction on Archbishop Ussher's[15] earlier estimate of 1056. His use of his other sources against this scale is thorough and ingenious: some of it reads like a detective story. Of course today we know his initial assumption was unsound and indeed try not to blush for him spending so much time on this sort of thing when he was doing so well on celestial mechanics. But Newton was interested in the whole riddle of nature and what he did in chronology was fairly part of that interest. Certainly he unquestioningly accepted certain things on authority: but so do we even today when most of us assent to the existence of electrons. Newton was in error: are we so sure that we are even aware of all our contemporary assumptions?

The biblical chronology was not shaken till in the early 1830s Lyell[16] convinced geologists of the uniformitarian doctrine: that all we find in the rocks can be accounted for by those same forces of destruction and formation that we see in operation today, provided we grant unlimited time for their operation. Promptly the geologists began to ask for enormous durations. Charles Darwin himself even proposed that 300 million years had elapsed since the chalk was laid down in the Weald.

Then abruptly a new figure entered the field. In 1866 Professor William Thomson, afterwards Lord Kelvin,[17] published a note entitled '*The Doctrine of Uniformity' in Geology briefly refuted.* The heat which he knew to be conducted out of the earth was so great that in former ages the earth must have been at a far higher temperature and must be cooling. The geologists could have 100 million years and not a day more.

T. H. Huxley[18] vigorously defended the geologists, though for all his skill he could not see what was wrong:

I do not presume to throw the slightest doubt upon the accuracy of any of the calculations made by such distinguished mathematicians as those who have made the suggestions I have cited. On the contrary, it is necessary to my argument to assume they are all correct. But I desire to point out that this seems to be one of the many cases in which the admitted accuracy of mathematical processes is allowed to throw a wholly inadmissible appearance of authority over the results obtained by them. Mathematics may be compared to a mill of exquisite workmanship, which grinds you stuff of any degree of fineness;

but nevertheless what you get out depends on what you put in; and as the grandest mill in the world will not extract wheat-flour from peascods, so pages of formulae will not get a definite result out of loose data.

But the forces of the Physical Establishment were strong and several remarkable 'yes-calculations' from the amount of sodium in the ocean and so on agreed strangely with Kelvin's estimate. And then one day we were told, so to speak: 'Delete one hundred and insert three thousand million because of radio-activity. Signed, E. Rutherford.' And the geologists said, 'I knew there was a catch.'

The whole of this controversy is of great interest because of the light it throws on the attitude to natural phenomena at different times and in different sciences. As we see it today, Huxley's answer to Thomson would have sufficed if he had said, 'My dear Thomson, you have omitted to allow for the possibility that much heat may be engendered by transmutation of the elements'. Such a remark at that time would indeed have caused an explosion. Both sides tacitly accepted that all sources of available energy had been displayed.

What is so clear from the history of hypotheses about the age of the earth is the extent to which at each stage these have depended upon contemporary assented assumptions, a set of assumptions which is not even the same for different groups of investigators. We also see how this set of assumptions has a strong emotional aspect as well as a logical one. According to our times and to our experience we represent the natural and the human world by a great set of images. To this set of images we apply, as a template, a system of hypotheses which seems to us coherent. The difficulty in scientific advance arises when some new experience necessitates a reassembling of the pattern of our images. There is immense resistance to this, for that is the basis of our common sense. Without it we are without guides or landmarks. Really fundamental departures from our accepted set of images require something in the human faculty of the highest quality; if we are to retain sanity it dare not be a total rejection of our traditions, but it must require what Samuel Taylor Coleridge[19] calls 'that willing suspension of disbelief for the moment'. Without it we cannot reassemble the images in a new and also coherent pattern.

About our image of the natural and human world we are willing to make assertions. John Henry Newman[20] in his *Grammar of Assent* expressed a view of this which agrees fully with my own experience from the contemplation of natural phenomena. Assent to such assertion is unconditional. It may be unconsciously taken. The propositions to which it is given may in fact be false; but it is given. Though ostensibly it may seem to be the result of conscious inference, it is in fact a response to many influences, rational and non-rational, conscious and unconscious, to what he calls the illative sense. It is not the mere logical recognition of an inference from premisses. Indeed a man convinced against his will is of the same opinion still. Newman had even gone so far as to say that 'man is *not* a reasoning animal; he is a seeing, feeling, contemplating, acting animal'. As I see it, the second half of that statement is true, but not the first. It is just because man has in fact two ways of arriving at a conclusion, the one rational and the other not, that he is distinguished from other animals—and from our present calculating machines.

I was first impressed with this duality when, as a naturalist, I became interested in the identification of certain obscure worms.[21] Anyone who does this cannot help being struck by the contrast between the way one identifies these animals in the museum and the way it is done in the field. Correct systematic identification in the museum depends on determining characteristic features. In my worms these are particularly internal features of the genital organs, of the musculature, and so on. These are, of course, invisible in the living animals. Nevertheless, I can unhesitatingly and accurately identify individuals of the various species of my worms in the field at several yards' range, even though they are only 10–20 mm long and 2–3 mm in breadth. But clearly, when in the field I make the affirmation, 'That is *Rhynchodemus bilineatus*', I am doing something quite different from what I do when I make this affirmation in the museum.

In the museum I seek a collection of definite 'yes or no' characters, those used in a key. As I have said, this sort of selection is not to do with the organism: it is to do with my logical processes. After we have selected the 'yes or no' characters, a very great deal of the impression which the organism makes upon us still remains 'unused'. This residue is undoubtedly important in our recognition of species even though it cannot be analysed in just this way. This residue, and indeed the whole impression made by the organism, is used when we recognise a species in the field. Field recognition cannot easily be translated into accurate analytical description. But it can be communicated to others. It can often be conveyed vividly by metaphor, simile and association, in fact by the ordinary modes of poetical expression: fiddler crab, old man's beard, glow-worm, and Queen Anne's Lace.

Aesthetic recognition and recognition by systematic deduction differ fundamentally. The first seems to be instantaneous; it deals with the whole impression of an object, and by it we apprehend many kinds of relationship at the same time. But it is subject to error. The second deals only with 'yes or no' characters extracted from the impression, and requires time for the successive deductions by which we reach our conclusion. But its conclusion is certain. This last difference between the two methods is important. The uncertainty of aesthetic recognition involves personal judgment; and this must be used with a quality which one can only call honesty. It is so easy to fill in the rest of the impression with what one's instant apprehension leads one to expect. At best that leads one to miss something important. At worst it enables one to reach the conclusion one would like for personal reasons.

The faculty for what I have termed aesthetic recognition seems identical with what Newman called the illative sense, that sense which allows spontaneous divination by the mind that a conclusion is true, which uses every kind of information experience has given us, and which uses it in a way different from logical inference. It works pre-eminently by instant analogy. Indeed some of its characters, including some of its errors, are comparable to those of analogue computing machines, in contrast to those digital computing machines which are able to give us deductively logical conclusions.

That the mind has two ways of arriving at a conclusion is in keeping with all that can be learnt about the evolution of behaviour in animals. Indeed the study of animal behaviour suggests that some features of our power of reason are much older than mankind. Whitehead[22] once said:

The first man who noticed the analogy between a group of seven fishes and a group of seven days made a notable advance in the history of thought. He was the first man who entertained a concept belonging to the science of pure mathematics.

But Köhler[23] has brought remarkable evidence that birds such as ravens and parrots not only have a 'prelinguistic' power to count, something of that power to extract number; they also seem to be able to do this in two ways, ways which recall the two ways we recognise an object—aesthetic, and by inference. They can be trained to expect suitable rewards by the recognition of a certain specified number of a variety of objects, however they are thrown together. The limit of this instant recognition can reach the number seven—just as it can in human beings. The other method of counting is even more remarkable. If I may quote Professor W. H. Thorpe[24] on Köhler's work:

It is as if the bird is doing some inward marking of the units he is acting upon. This supposition is strengthened by the fact that sometimes these supposed inward markings show themselves in external behaviour in the form of intention movements. Thus a jackdaw, given the task of raising lids until 5 baits had been secured (which in this case were distributed in the first five boxes in the order, 1, 2, 1, 0, 1) went home to its cage after having opened 3 lids only and consequently having eaten only 4 baits. The experimenter was in the act of recording 'One too few. Incorrect solution', when the jackdaw returned. It then went through a remarkable performance: it bowed its head once before the first box it had emptied, made two bows before the second box, one before the third. It then went further along the line, opened the fourth lid (the box which had no bait) and then the fifth and took out of this the last, fifth bait. Having done this it left the rest of the line of boxes untouched and went back to its cage as if regarding the experiment as over. It appears from this intentioned bowing repeated the same number of times before each opened box as on the first occasion when it found the baits in them that the bird remembered its previous actions. It seems as if it became aware that the task was unfinished and so returned and commenced again picking up *in vacuo*, with intention movements, baits which it had actually picked up. When, however, it came to the last two boxes which by mistake it had omitted to open on its first trip, it performed the full movements and thus completed the task. The simplest explanation Köhler has to offer for this intention marking is that it may consist in equal marks—as if we were to think of or give one nod of the head for one, two for two, and so on. This is called thinking unnamed numbers...

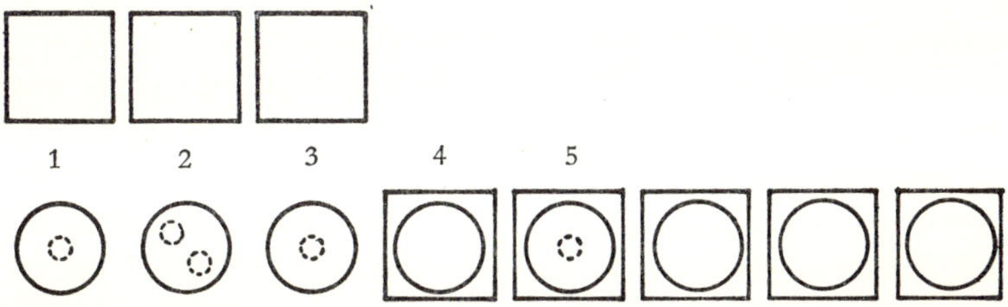

Figure 18. Evidence for conclusion that a bird does some inward marking of the units it is acting upon when demonstrating its 'number sense' in an experiment. The figure shows the state of affairs when the jackdaw has removed the cards covering the first three boxes and has taken the four baits which they contain. The remaining boxes are closed; number 4 is empty and number 5 contains the last 'allowed' bait. After O. Köhler, 1950. For further explanation see text.

So often, as soon as we seem to grasp some absolute characteristic of the human faculty we find it foreshadowed in much simpler creatures. That does not mean that the gap between us and them is not immense: it means again that the simpler qualities are those our minds most easily grasp.[25]

Aesthetic recognition that an animal belongs to a certain species is one aspect of the exercise of Newman's 'illative sense'. Such recognition forces itself upon our minds.

Both the recognition of a species or of some significant natural event are the first stages of a scientific investigation. We may end by publishing a paper about the matter. When we do there is little evidence of the route by which we apprehend the significance of the phenomenon.

All this is equally true of an experimental investigation. When properly displayed in publication both start from clear premisses and arrive logically at a conclusion. Often this looks so straightforward one is surprised that no one had done it before.

But except in those researches which are mere vain repetition for the purpose of increasing the volume of personal publication or which are doing no more than following out some almost forgotten conclusion, the published logical sequence often bears little relation to the route by which the research was initiated. Authority, analogy, and every aspect of the illative sense play their part. There are often quite strong emotional drives, the goals of which are not apparent to the investigator.

The origin of the perception of significance in a phenomenon is the system of images by which we represent the natural world. That in turn is based upon the models and analogies of our childhood and youth. For me, that is roughly the period from 1900 to 1920. The constructional toy 'Meccano' was invented just when I was of an age to use it. The images of telegraph and telephone circuits were in the minds of my contemporaries. We had no electronics. But a boy could buy a Wimshurst machine for twelve and six without coming into collision with the law which protects television addicts. And those were the days when non-ferrous metals and soldering irons were cheap.

Each of us grows up with such a system of models. As natural science advances we have to patch up the system of images built in our youth with new knowledge: we try to keep up with the young Dr Joneses. Some do this well and some badly. But even when it is well done each new generation brought up with new models has an advantage—as every generation has found. Sometimes in my work I have to use electronic devices. I am not handy with them. What is more, I do not think in terms of them. Many of my young friends do just that and can swiftly perceive relationships as well as ways of doing things that are not within my easy grasp. I tend to think and devise in terms of sealing-wax, solder, and $C = E/R$.

Analysis of how a piece of research begins is such a personal affair, as is anything which begins with the operation of the illative sense, that it is best if I indulge in reminiscence—to add just another personal account to those of so many others.

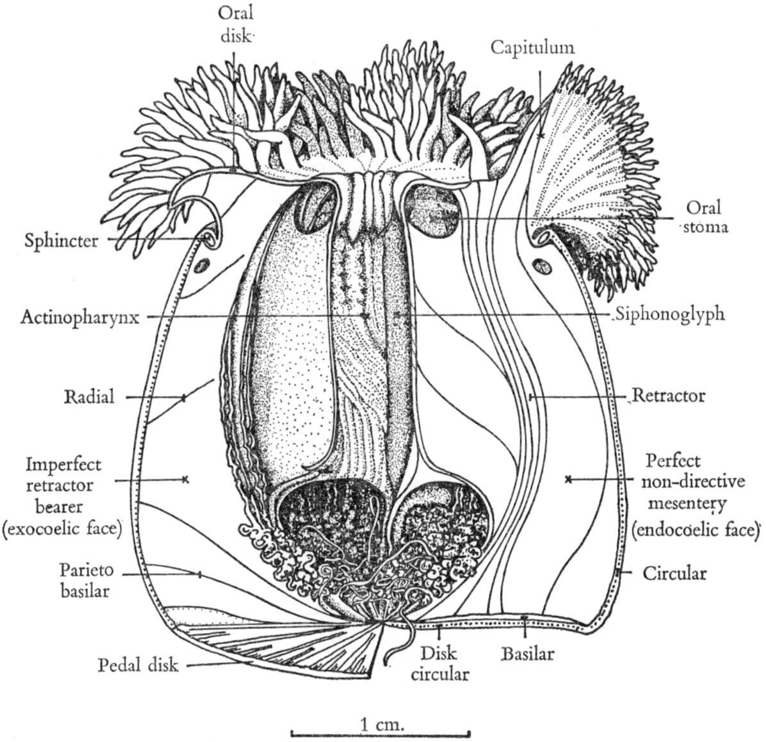

Figure 19. Diagram of the general muscular organisation and anatomical structure of the sea-anemone *Metridium*. (From E. J. Batham and C. F. A. Pantin, 1951, The organisation of the muscular system of *Metridium senile*, *Quart. J. Micr. Sci.* **92**, 27–54.)

Some thirty years ago I made what proved to be a rather interesting discovery about what one might call the machinery of behaviour in the simplest kinds of cellular animals: sea-anemones. Their bodies are fairly simple muscular sacs. But they can feed, and when you prod them they shut up by reflex action. For a long time it was thought that they had no nervous system. Then it was found that they did possess a network of nerve fibres—but they certainly have no brain. Yet, like the higher animals, they show a great variety of purposive movements. How can that be brought about by such a very simple nervous system?

I have said that it had been shown in the higher animals that excitation is transmitted in nerve fibres by impulses which are independent of the strength of an electrical stimulus. Earlier workers on sea-anemones concluded that if such a simple network of nerves conducted 'all or nothing' impulses the anemones ought to be capable of only one kind of reflex response. But since in fact they could do all sorts of things there must be some different and more primitive mechanism of transmission of excitatation to give varied responses in the muscles.

In the published account of my experiments[26] I showed that, just as the existence of nerve cells would lead one to expect, there was a system which conducted excitation in these animals, and that this excitation has all the properties of the nervous impulse found in the nerves of higher animals. I began by proving that the class of phenomenon to which the nervous impulse belonged in sea-anemones was identical with what Keith Lucas and Adrian[27] had shown to be true for our nerves. I then gave evidence that the varied movements of which the animal was capable were due to a valve-like control of the transmission of nervous impulses from one nerve cell to another, and from the nerve cells to the muscles. I then showed that the whole system was planned rather like a very simple disseminated brain.

Now the actual order in which my ideas developed and the order in which the experiments were done was quite different from this. The course of the research shows quite clearly the importance of authority and contemporary fashion in directing one's attention to phenomena. It also shows the importance of the illative sense in reaching conclusions. It shows the importance of contemporary models and the importance of the aesthetic and emotional aspect in initiating research. Moreover, the development of the work underlines the importance of chance—though the element of chance is that with which the scrum-half is familiar rather than that of the winner in Premium Bonds.

When I began to read about general invertebrate physiology I found that by vertebrate standards the phenomena were full of anomalies. Some of the explanations offended my sense of elegance. It was commonly assumed that the 'all or nothing' rule of nervous conduction was something that could be evolved gradually and by degrees. That seemed to me absurd; a class of phenomena cannot be gradually evolved.

My research on sea-anemones began in the following way. I had spent several months at the Naples Biological Station studying nervous conduction in crabs. I had shown to my own satisfaction that, notwithstanding the common view, conduction in crab nerves was of the same class as that in the nerves of vertebrates. I also found that the response of the muscle was greatly influenced by the number and frequency of electric shocks applied to the nerve. This influence was apparently controlled by a valve-like property of the junction of the nerve and the muscle.

To do this work I was using a simple method of repetitive electrical stimulation by electrical condensers of different sizes. In the late spring I had carried on this work on crabs to a natural halting place, and looked round for something to study during remaining weeks at Naples. It so happened that the previous occupant of my room left some sea-anemones in the aquarium in which I kept my crabs. I thought it would be fun to repeat some old experiments in which one partially separated a long slip of tissue from the body of an anemone, so that by giving a shock at one end I might measure the exceedingly slow nervous conduction which was supposed to take place. I prepared an anemone in this way and I found that electrical excitation was quite easy, and could be detected by noting the closure reflex of the body of the animal when at last excitation reached it.

At first I sent in bursts of electric shocks at the rate of about ten to twenty per second—just as one would do for a crab or a vertebrate—and I noted with real delight the long time that elapsed before the excitation reached the body of the anemone and it shut up.

At that time, as I have said earlier, a good deal of attention was paid by physiologists to the optimal duration of an electric shock for the excitation of a nerve. A shock that is too brief shows a higher threshold for excitation. On my anemones I found that to get the best results I needed long slow shocks. I gave these by means of very large condensers which took some time to discharge.

Owing to the limitations of my apparatus I could only send in such long slow shocks at quite low frequencies—at about one per second. By happy accident I found that when I did this, each shock of the series caused the anemone to give a discrete contraction. This allowed me very easily to measure the slow speed of conduction of excitation.

Under the influence of the contemporary idea of the importance of the duration of an electric shock, using a standard frequency of stimulation of one per second, I noted the threshold of intensity when I varied the size of the condenser to give shorter or longer shocks. I got quite good curves relating threshold to condenser size. But all this had really little to do with the important phenomenon I had hit upon—and which I had not yet appreciated.

In my next experiments I stimulated the tentacles and mouth region of the animal with shocks. I then found that an appropriate series of electric shocks could make the anemone swallow the electrodes as though they were food. I found that the extent of the effect depended particularly on the number and frequency of the shocks given.

The contemporary view was that unlike 'all or nothing' conduction in vertebrate nerve, in simple creatures like sea-anemones the excitation gradually died away as it spread from the stimulus, as for instance when that was applied to the mouth region of an anemone. I saw at once that the extent and size of the response depended simply on the number and frequency of shocks applied.

My attention having been called to the importance of the frequency, I at last saw that most of my previous experiments were more fascinating than they were important, and that what I should do was to apply shocks of known intensity, frequency and number to the body of an intact animal and to record systematically what were the conditions which made the animal shut up.

As soon as I did this, I found at once that just as in the nerves of vertebrates the effect of the electric shocks was independent of the strength of the stimulus: excitation was quite beautifully 'all or nothing'. The size and nature of the reflex response were controlled by the frequency and number of the shocks. These at last were of course the cardinal experiments upon which the logical argument of any published paper should be based.

I thus arrived at the final ordered argument by a singularly roundabout route. Nevertheless, although my first experiments seemed misdirected, I knew very well that somehow or other I had hooked a big fish. I happened at that time to be in a state of increasing ill-health with great fatigue and lassitude. Nevertheless, the interest and excitement were such that I worked at great speed for long hours: these initial experiments were conducted in five days, and three days later I had done enough to see all the implications recorded in my first published papers. Of course after that there was a long period of repetition and improvement of technique to give more elegant representation of the phenomena. During this period new notions and extended implications would occasionally suddenly present themselves.

What I have said here about my own experience is, I believe, not unlike the experience of many others in the detection of new ideas and new significant phenomena. The springs of action which drive one to an investigation are many, and many of these have a strong emotional content. Intuitive perception is an essential preliminary step. There are many instances, from Newton to Poincaré,[28] that this is true in the physical sciences as well as in biology, and notwithstanding the opinion of Dr Agnes Arber,[29] I agree with Poincaré on the importance of a change of occupation—or of sleep—in providing a condition for the spontaneous intuitive solution of a problem. But I think this intuition does more than this. It does not only suddenly present solutions to our conscious mind, it also includes the uncanny power that somehow we know that a particular set of phenomena or a particular set of notions are highly significant: and we are aware of that long before we can say what that significance is. I think myself that this is related in some way to the intuitive detection of conformity, unconformity or far-reaching extension of that accepted system of images by which we represent the natural world in our conscious minds. Change of occupation, or the forcible apprehension of some phenomenon from a novel point of view, can suddenly engender that 'willing suspension of disbelief for the moment' necessary for the reassembling of that image. We cannot name it, but we become aware that the image is undergoing metamorphosis even though we cannot yet tell what the image will become.

I do not think the restricted and the unrestricted sciences differ in principle in respect to these conclusions. But in the richness and variety of the phenomena presented in the unrestricted sciences and the short half-life of their hypotheses we are more frequently thrust back upon intuition to initiate our work, just as for those same reasons we are less likely to be guaranteed a long safe-conduct through the magnificent intellectual sequences of the mathematical models so familiar to the physicist.

Summary

The argument of these Tarner Lectures shortly set forth

NATURAL PHENOMENA involve significant objects and events at many different levels of size and complexity. The questions which the scientist tries to answer are at all these levels, and the immediate objective at any of these is complete in itself. Investigation is not just analysis indefinitely extended to smaller and briefer things, or by smaller and briefer measurements.

When we view the different sciences, it is fair to distinguish those that are restricted from those that are necessarily unrestricted. The physical sciences are restricted because they discard from the outset a very great deal of the rich variety of natural phenomena. By selecting simple systems for examination they introduce a systematic bias into any picture of the natural world which is based on them. On the other hand, that selection enables them to make rapid and spectacular progress with the help of mathematical models of a high intellectual quality. In contrast, advance in the unrestricted sciences, as in geology and biology, is necessarily slower. There are so many variables that their hypotheses need continual revision.

When studying any of these natural phenomena the scientist is seeking for relationships. In fact, relationships are among the things we directly perceive and on which our conviction of the reality of the material world is based. In calling them relationships we must not suppose that our perception of them is a matter of inference. It is direct and in a great measure communicable to others.

The real world presented to us in this way consists particularly of enduring objects and equilibria on whose continued presence our common sense can rely as a frame to which events can be referred. For all the complex development of hypotheses about them, the objects and events which concern the physicist are simple. But even in the physical world there are complex systems, like those studied by geologists and meteorologists, and the machines of the engineer. Order and complexity, absent in simple systems, must inevitably arise through the operation of the second law of thermodynamics and the principle of Le Chatelier, given the remarkable and restricted properties which we find material objects to possess in this universe.

Living organisms seem at first sight to have physical qualities which distinguish them absolutely from inanimate objects. In former days 'vital forces' were invoked to account for their special features and for what organisms do. This distinction arises partly because the physicist's imagery tends to be based upon simple physical systems. Complex inanimate systems in fact show many of the features we associate with living things. All the same, living things show certain features in a transcendent degree, particularly purposive organisation and behaviour and their power of reproduction. But, in all they do, they do not disobey the physical and chemical generalisations derived from the study of inanimate systems. Given that, as material configurations, systems such as the living organism are possible, the operation of the laws of thermodynamics will inevitably tend to the preservation of such systems; that is, particularly through the operation of natural selection, which is itself one of the laws of thermodynamics governing the fate of matter and energy in complex systems.

The difficult question is how the complex purposive material configuration which composes a living organism could arise. Apart from the general question of the origin of life itself we are still faced with that of how a living organism could arise from the germ from which it springs in each generation. Mechanistic analysis has shown surprisingly how in fact that is possible through the unique properties of a particular molecular machine, the class of nucleic acid DNA, and the unique sequences of events which follow its operation in each organism. But that this is so is dependent on the emergent properties of this molecular configuration.

Like the other laws of thermodynamics, natural selection only tells us in which direction systems will develop. It tells us nothing of the unique properties of particular material configurations and of the events associated with them. All it requires is that any system must meet such and such a general specification if its probability of survival is to be high.

Of many of the unique emergent properties of complex systems we have no explanation in terms of simpler ones, though that represents our empirical state of knowledge rather than a proof that we must add to the assumptions necessary to describe simple systems if we are also to describe more complex ones. In any case all such assumptions apply only to the contemporary template of hypotheses which we apply to reality rather than to the real world itself.

The objective of natural science is to discover the relations of things and events at different levels of complexity. That is, it is concerned with the principles underlying classification. Classification is inherent in perception, which operates through direct apprehension in the light of past experience of that which we may subsequently choose to analyse logically as relationships.

Because of their complexity it might be supposed that living things would be the most difficult natural objects to classify. This is not in fact so. Classificatory keys for the identification of species are easily made; but experience also shows the existence of a so-called 'natural' classification of organisms as well. The possibility of constructing keys reminds us of the deductive logical basis of our classifications: they are convenient to our mental processes. The empirically discovered 'natural' classification in part displays relationships of all organisms. This classification has a linear-branching character. We have good reason to suppose that this is a consequence of the linear-branching course of evolution. By a happy accident, this supplies a system that is amenable to logical analysis, so that this 'natural' classification can be called so partly because it is natural to our minds. It does not in fact perform all the duties of a complete natural classification, that is, it does not display all the relations between organisms which the scientist discerns.

For a logical hierarchical classification we need (*a*) precise definition of objects, and (*b*) linear-branching relationships of objects. The units of the physical world, molecules, atoms and ultimate particles, admit precise definition much better than do species of organisms. But these physical objects exhibit relationships which can only be represented in a multidimensional system: a 'species' of them may fall into more than one genus—there can be no linear-branching arrangement, a fact which creates havoc in chemical nomenclature.

The 'natural' classification of species of living organisms allows linear-branching arrangement. Does it allow precision of definition? As William Whewell pointed out, species are not given by definition but by type. They are determined by a point within, not by a fixed boundary without. But it is important to remember that this type by which we recognise a species or any other classificatory grade is not a material specimen. It is in the trained machinery of our perception, trained by the integration of all our past experience of such objects and their relationships.

Among the relations not displayed by the so-called 'natural' classification of organisms are those concerned with functional analogy, such as the resemblance of the eye of an octopus to our own and to a photographic camera. Such resemblances may be very detailed. There is in fact a system of classes of engineering solutions of particular functional problems. Because of the hierarchical relationship of functional parts to the whole organism a functional classification of organs is possible which exhibits a linear-branching character. This is quite distinct from the 'natural' classification and can supply the basis for lecture or chapter headings in functional physiology.

But apart from the classification of objects there is also a classification which should display all the relationships between events concerned with those objects. Such classes are those in which events display common physical principles of action—as in relaxation oscillations. This is a multidimensional classification. It is of great importance because the mental image of these classes supplies the models by the operation of which, directly or in the imagination, we seek to understand and explain phenomena. Determination of the class to which an object and its associated events belong is an essential feature of the scientific method.

Scientific research begins with the apprehension of some notion or the perception of the significance of some phenomenon in relation to our image of the natural world. This demands confirmation, extension, or even reorganisation both of the image and of the template of hypotheses by which we purport to explain it. From our old and new hypotheses we deduce consequences which can be verified by selected observation or experiment. In extending our hypotheses we observe the principle of economy of hypothesis. We go further: faced with the overwhelming variety of nature, we select what seem to us to be the simplest phenomena for investigation. That may enable us to erect mathematical explanatory models of a very high intellectual order. But thereby we tend systematically to exclude complex problems from investigation.

The complex problems of the unrestricted sciences are much more difficult than those of the restricted ones. Their hypotheses continually break down. Moreover, they present special difficulties of attack, particularly in the biological sciences, associated with the operation of natural selection. This can lead to specious simplicity in the apparent relation of cause and effect—particularly in the analysis of behaviour. Also it can undermine the simple experimental approach to the determination of the relation of cause and effect. These difficulties can be demonstrated by the study of the history of our hypotheses about nervous action.

But these matters all concern the development of our scientific attack once the problem itself has been posed. That problem arises in some way from the system of images by which we represent the world in our minds. That system and our template of hypotheses are profoundly influenced by that to which we assent on authority—which may be wrong. It is also influenced by models based upon our experience of the natural world in our youth and on our contemporary enthusiasms. We perceive the problems to be attacked and first see where they will lead through what John Henry Newman calls the illative sense, by which we arrive at a conclusion from all manner of information, much of which is not of the kind suitable for logical deduction. The illative sense and its relation to logical verification are clearly seen when we compare the recognition of a species of organism in the field with our recognition of it by taxonomic identification in the laboratory. This dual method of arriving at a conclusion is an essential part of the scientific method in experiment also. Moreover, there is evidence even in the behaviour of animals other than man of a dual method of recognition.

It is a remarkable feature both of our minds and of the world we experience that two such very different ways of arriving at a conclusion as that of the illative sense and that of logical deduction should so often arrive at the same result. There is a story of how Newton informed Halley of one of his most fundamental discoveries of planetary motion. 'Yes,' replied Halley, 'but how do you know that? Have you proved it?' Newton was taken aback—'Why, I've known it for years,' he replied. 'If you'll give me a few days, I'll certainly find you a proof of it'—as in due course he did.

When we examine an actual experience of the genesis of a problem in scientific research all these factors are evident: the authoritative background, prediction by the illative sense, and the selection of features for logical verification. Such a study shows how very different is the final logical exposition of research in the published work from the route by which the problem was first approached. In that, the illative sense plays an essential part, not only by pointing to a conclusion but also by pointing to a notion or phenomenon as significant—even before the significance is understood. It initiates a powerful emotional drive towards the solution of the problem, perhaps through the unconscious perception that some part or all of that system of images by which we represent the natural world to our minds, and which is the basis of our common sense, is about to undergo metamorphosis.

APPENDIX I

Life and the conditions of existence

PERHAPS THE MOST remarkable conclusion about the universe presented to us today by astronomers is that notwithstanding its vast scale and the strange notions about space and time which we are forced to adopt, even those galaxies the most remote from ourselves in distance and in time are apparently composed of those same elements from the same periodic table, and so far as we can see are aggregated into the same kinds of configuration and subject to the same relations to energy as those with which we are familiar in our solar system.

Knowledge of this universality of the properties of matter and energy is based upon the behaviour of inanimate materials. But living organisms also are set in a material universe and are built of the same kinds of material particle; and it has become abundantly clear that these same laws of matter and energy hold for the living organisms and indeed for our own bodies.

This does not mean that the laws deduced from the study of inanimate systems will necessarily serve to explain the special features which distinguish living matter, or the behaviour of an amoeba or of a mammal, or the expressions of the human faculty. So far, we only say that the manifestations of none of these special attributes require the contravention of those physical laws, whatever additional ones may be required for their description. But it is of interest to consider how far the origin and nature of life and of the human faculty are determined by the conditions of existence as we know them. It is with this that this essay is concerned. To deal with so vast a subject requires apology; neither a biologist nor anyone else can pretend to adequate knowledge over more than parts of it. But that is no reason to leave it undiscussed.

It became apparent in the beginning of the nineteenth century that, peculiar as they might seem, the substances of living organisms obeyed the same chemistry as did those of the inorganic world. But their organisation showed certain features which were unique among natural objects. Those were: the adaptation of the parts to the functions they performed and the fact that at least in the animal world the anatomy was based upon one or another of a restricted number of types, that of the vertebrate, that of the insect, and so on. It was this that gave rise to the concept of 'homologous organs': organs based on the same plan but used for a different function, as in the hand of a mole and the wing of a bat. Relationships of this sort are not obvious in the inorganic world.

Like many others, Richard Owen, in 1848, interpreted both adaptation and the underlying plan as evidence of divine design. Agreeing with Paley he says:

> With regard to the adaptive force...its effects must ever impress the rightly constituted mind with the conviction that in every species ends are obtained and the interests of the animal promoted, in a way that indicates superior design, intelligence and foresight in which the judgement and reflection of the animal never were concerned; and which...we must ascribe to the Sovereign of the Universe...[1]

And the homologies of the underlying plan we must surely consider as '...manifestations of some higher type of organic conformity on which it has pleased the divine Architect to build up certain of his diversified living works'.[1]

In 1859 Charles Darwin offered a different explanation of these anatomical features: 'For natural selection acts by either now adapting the varying parts of each being to its organic and inorganic conditions of life; or by having adapted them during past periods of time...'[2]

Adaptation was the inevitable consequence of selection: unity of type was attained through the inheritance of former variations and adaptations. There was no need to invoke a designer to account for the phenomena; time and chance sufficed.

Darwin when defining natural selection directs attention to the importance of environment, and indeed of the general conditions of existence, in determining the special characteristics of living organisms: 'Let it also be borne in mind how infinitely complex and close-fitting are the mutual relations of all organic beings to each other and to their physical conditions of life.'[2]

The force and direction of natural selection are determined by the environment and by the other conditions of existence. Indeed, the nature of variation itself is determined by the properties of matter included in the conditions of existence. Yet, for all the vast increase in our knowledge of the genetical theory of evolution, there seems to have been surprisingly little analysis of this essential element in the theory of selection and of adaptation. Even in so thorough a survey as Huxley's *Evolution, the Modern Synthesis*, 'environment' is not to be found in the index.

As R. A. Fisher says in his *Genetical Theory of Natural Selection*: 'Now, survival value is measured by the frequency with which certain events, such as death or reproduction, occur to different sorts of organisms exposed to the different chances of the same environment...'[3]

But this must not lead us to assume that the environment is some simple constant thing providing a natural asymptote towards which adaptation of the species is forced to approach. The selecting environment is itself complex. Indeed this was realized by the old natural theologians who not only saw design in the conies but also in the stony places which had been made for their habitation.

In the first place, the environment itself has organismal qualities. In a manner which I shall discuss later it may even show 'homoeostatic devices'; those 'built-in' devices which in living organisms so characteristically maintain constancy when external conditions change.[4] As in living organisms, we find configurations which repeat themselves, as in deserts, lakes, forests, shores and so on. Particular examples of these configurations can, during geological time, grow and divide. Their reproduction is, however, simple. There is no heritable variation upon which natural selection can act as in living organisms, except in a complex way consequent upon living things being among the environmental components. Environments can become extinct through climatic or geographical change. Such changes can also lead to their spontaneous generation in new regions.

These configurations provide environments for species of animals, plants and bacteria. Their history is of profound evolutionary importance. Thus, one class of these configurations includes fresh-water swamps and rivers liable to stagnation and desiccation. The existence of these in Devonian times was accompanied by the rapid evolution of the Dipnoi or lung-fishes. Today these fishes are still to be found, but little different in form from their remote ancestors; and we even find fossil records of burrows such as their modern representatives form under adverse conditions of drought. The lung-fishes are well adapted to their special environment. In Uganda they are amongst the common marketable fishes of the villages. But today they are restricted to central Africa, central South America, and the south-east of Australia. In these situations we can say that, just as the Dipnoi of today represent the little-changed descendants of those which inhabited an ancient environment in which they were evolved, so also that remote kind of environment has itself survived to the present day. There has been no break during which the Dipnoian species occupied a totally different sort of habitat. In many parts of the world once inhabited by Dipnoi geological evidence shows that their environment became totally extinct—as in Devonian rocks of Scotland—and the animals have died out. Elsewhere in the world we find other swamps in which there are no Dipnoi but in which other very different kinds of fish, notably the siluroids, have developed the power to breathe air. There is no reason to suppose that these are more efficiently adapted to this kind of environment than the Dipnoi; but the Dipnoi could only inhabit an appropriate environment provided this has had a continuous connection in time with the ancestral environment which selected them.

The history of the Dipnoi reminds us that the evolution, extinction and reappearance of a particular kind of environment is a major factor controlling the origin of species. A species may become extinct, not because of the presence of a more efficient competitor but because the environment to which it was adapted has totally failed to survive.

But we must not identify the total sum of physical and biological factors which comprise a particular environment with that which is enforcing adaptation. A certain minute terrestrial worm, *Geonemertes novaezealandiae*, feeds on the remains of minute insects in woodlands in New Zealand. It is to be found in several quite distinct plant-associations; as in tree-fern forests, in the broom-like *Leptospermum* thickets, and in the cold forests of the southern beech, *Nothofagus*, below the snows of the Southern Alps. Manifestly, only certain of the various physical and biological features which characterise, say, the tree-fern forests are of essential significance to this worm, and not to others. 'The same environment' in Fisher's measure of survival value is by no means a simple concept.

I have chosen these examples to illustrate the fact that the environment which a particular species locally inhabits is itself a member of a class of configurations which have organised properties and an evolutionary history of their own; and that only certain features of a configuration are essential for the local existence of a species. It is characteristic of these environmental configurations that they offer necessary conditions for the existence of the species. And it is characteristic of the universe as we know it that configurations offering these conditions can exist.

The necessary conditions for existence of an organism are not simply the presence of a particular external environment, though the phrase is sometimes used in that sense. They include certain characteristics of that environment, but they also include characteristics of the materials which compose the organism itself. This is the sense of the phrase when used by Cuvier: 'Comme rien ne peut exister s'il ne réunit les conditions qui rendent son existence possible, les différentes parties de chaque être doivent être coordonnées de manière à rendre possible l'être total, non seulement en lui-même, mais dans ses rapports avec ceux qui l'entourent...'[5]

These conditions not only include the necessary functional relationship of bones and muscles. Equally it may be said that among the conditions for the existence of life as we know it are the properties of carbon: and these include not only those which give the stable alkalinity of the environing ocean through the presence of bicarbonates, but also those which concern the essential molecular structure of the organism itself. When we speak of the adaptation of the organism to a particular environment we are in fact concerned with both internal and external conditions at many structural levels which permit it to exist. Indeed, we cannot really separate an organism from its environment. It is not merely the impossibility of defining a precise physical boundary, but all we perceive about an organism involves the world external to it. We are, in fact, very much in the position assumed by Michael Faraday in his attack on Dalton's atomic theory: 'What thought remains on which to hang the imagination of an *a* independent of the acknowledged forces?'[6] He then goes on:

> The view now stated of the constitution of matter would seem to involve necessarily the conclusion that matter fills all space, or at least all space to which gravitation extends (including the sun and its system), for gravitation is a property of matter dependent on a certain force, and it is this force which constitutes the matter. In that view matter is not merely mutually penetrable; but each atom extends, so to say, throughout the whole of the solar system, yet always retaining its own centre of force.

Let us examine the conditions of existence by surveying the nature of the material world as we experience it. The things we say we perceive in the world have two components. These are enduring objects of various grades of complexity: atoms, molecules, chairs, hills and so on. They are the residue of things liable to sudden or to gradual destruction. There are also 'open steady states' which require energy as well as matter for their maintenance: flames and rivers and living organisms are such.

The various objects are at various levels of complexity. Starting arbitrarily with molecules, we can go downwards to atoms; from these to the protons, electrons, mesons and the ever-increasing family of 'ultimate' particles into which modern physics analyses what we once considered to be indivisible units. We can also go upwards to higher orders of structure: crystals, thunderstorms, rivers, organisms and so on.

Within these orders of structure the molecule stands at an important boundary. Two molecules of the same substance, two hydrogen atoms, or two electrons, are indistinguishable from each other by individual features. There are no individual differences, and the species is absolutely defined. Objects of higher order than this, such as chairs, hills or living organisms, each have individual characteristics. They fall into species, but the boundaries of the species are not sharply defined.

There is a curious difference here between the classification of molecules, and simpler entities, and that of the higher organisms. In the former the species is absolutely defined, but all higher classes are complex; the same 'species', copper, can be classed in more than one 'genus' according to the way we view its relations to other elements in the periodic table. In contrast, species of animals and plants do not have precise boundaries, but because of their branching evolutionary origin in time, they readily admit classification into genera and higher categories. Species of molecules fall into higher classificatory units. But unlike living organisms the classification is multidimensional: that is, the same species can fall into more than one genus.

Crystals stand in an ambiguous position in that individual features of two crystals of the same species are indistinguishable except by features derived from their surfaces. Moreover, like living organisms, but for quite different reasons, the same 'species' does not fall into more than one genus.

But the most important difference between higher orders of structure than the molecule and those below is that the structures of higher order undergo denudation and wear out, whether they are hills or men. In contrast, the individual molecule or atom remains unchanged unless and until it undergoes destruction. This feature is particularly important for living organisms as we know them, for though in the adult stage they are higher-order systems which undergo attrition, during reproduction they pass through a stage in which future individual development is coded in the molecular organisation of nucleic acid molecules. Because of the indistinguishable character of their atomic components in molecules, essential individual characteristics can be reproduced unchanged in a manner impossible in gross material objects.

We thus get a picture of the natural world in which there are structures of increasing order of complexity, and in which there is an important transition as we pass from the molecular to configurations of higher order. At each level of structure special properties appear. Appropriate molecules arranged in an appropriate way can have the properties of a petrol engine, a computing machine, or a living organism. These properties relate to the configuration rather than to properties of the individual atoms and molecules of which it is composed. It is the configuration which gives rise to the computing machine rather than the copper atoms which happen to be used in its electrical connections.

These new properties which appear in particular configurations are conveniently called 'emergent'; and the most striking features, such as the living state, behaviour and the human faculty itself, seem to be properties of this class.

G. H. Lewes defined emergence as follows:

> All quantitative relations are componental, all qualitative relations elemental. The combinations of the first issue in Resultants, which may be analytically displayed; the combinations of the other issue in Emergents which cannot be seen in the elements nor deduced from them. A number is seen to be the sum of its units; a direction of movement is seen to be the line which would be occupied by the body if each of the incident forces had successively acted on it during an infinitesimal time. But a chemical or vital product is a combination of elements which cannot be seen in the elements. It emerges from them as a new phenomenon.[7]

It will be seen that emergents are not properties of the natural world itself, but rather of the hypotheses by which we perceive and represent them. To an uninstructed schoolchild, the theorem of Pythagoras might illustrate an empirically discovered emergent property of concrete right-angled triangles. But we would find that in fact it was implicit in the original axioms and postulates of Euclid. On the other hand, contemporary hypotheses about simpler systems may often be demonstrably inadequate for the explanation of the features of complex ones. Van der Waal's equation

$$\left(P+\frac{a}{v^2}\right)(V-b) = RT$$

served to represent many of the essential features of gases. But the assumptions of the kinetic theory of gases, even with the addition of the factor *a* representing attraction between the molecules, are insufficient for us to predict the features of the liquid and the solid states. Yet there seems no way in which we could decide that, however closely we examine the phenomena of gases, we could never at any future time derive a set of assumptions from the study of gases alone from which could be derived the emergent features of liquids and solids.

Empirically, the appearance of emergent properties is a striking feature of successful hypotheses about systems of increasing complexity. The possibility of absolutely emergent properties is uncertain. Nevertheless, examination of our hypotheses about the natural world does suggest that its events can only be represented by means of several independent assumptions. That is most clear when we consider the changes of matter and energy in time. The changes which experience shows us to be undergone by matter and energy are found to be governed by a series of empirical 'laws': those of thermodynamics.

There is (1) a primary law which states that things in thermal equilibrium with the same thing are in thermal equilibrium with one another; (2) the principle of the conservation of energy; (3) the so-called 'second law', which in its various forms implies that disorder tends to increase; and (4) there is Nernst's law, that by no series of finite processes is absolute zero attainable. Each of these laws involves independent assumptions. We could conceive of any one not holding.

Such 'laws' are in fact applicable strictly to ideal fully circumscribed systems. But this does not release us from applying them to the everyday world of our experience. They were in fact derived from experimental data in that world; and, apart from seeming anomalies due to misconception, events in that everyday world always fail to necessitate their contradiction.

Though derived from experimental systems with comparatively simple components, these laws are found to apply to complex ones, including living systems. If indeed organisms in fact succeeded in evading the so-called 'second law'—just as, for that matter, if animals succeeded in acquiring 'extra-sensory perception'—they would be at so considerable a selective advantage that no creature on earth could afford to be without such conveniences to their way of life.

It will be noted that the laws of thermodynamics do not tell us precisely into what particular pattern a configuration of matter and energy will pass with the lapse of time. They only set prescribed conditions which must limit the character of any change from an initial state. They do not tell us what route these changes will follow and how long their journey will take. Thus, if we are to predict what will happen when a hot material body is placed in conjunction with a cold one, we must know a good deal more than the 'second law' of thermodynamics. We must know a great deal about the distribution and physical properties of the material bodies; and that knowledge is ultimately derived from empirical laws which are quite independent of those of thermodynamics. When energy is introduced into a material system of even moderate complexity there may be many different pathways for it to follow and many different forms which it can take. Even so simple a system as two masses of moist air at different temperatures does not usually result in the simple, gradual diffusion of heat throughout the combined mass. It may even result in a turbulent system causing a lightning flash in which one small part of the system is raised to a vastly higher temperature than that of the original.

These considerations are of particular importance when we come to consider the nature of living organisms. In the gross 'real world' about us we can discern two classes of objects: on the one hand there are enduring fixed configurations, mountains, tables and so on, which undergo denudation but do so at a sufficiently slow rate for us to apprehend their individuality (with or without the aid of instruments). On the other there are 'steady states', particularly 'open steady states'. These may be apprehended as enduring objects, but they are not fixed material configurations. As in a whirlpool, for all their seeming constancy, through them there is a continual flow of matter and energy in which input balances output. This balance may be continuous if deviations cause the system to swing back to a steady mean; or it may be oscillatory, as in relaxation oscillations, where the system changes in a cyclical manner, but continually passes through a succession of states with the same properties. Any such systems may include some fixed material, as a burning gas jet.

Such enduring but dynamic systems include living organisms, thunderstorms, rivers, the flames of Bunsen burners, and many running machines of the engineer. Over a period long enough for us to apprehend them as objects, the input of matter and energy in these dynamic systems is balanced by the output. In the long run, denudation, gradual or calamitous, affects them all: men grow old, the necessary conditions for thunderstorms pass away, and every businessman annually writes off a percentage for the depreciation of his machines. But the very existence of these steady states depends upon the maintenance of the supply of matter and energy.

Living organisms have a feature which seems to distinguish them from simple physical objects. They have parts to which a future-directed function can be assigned, and given one of these parts we can make predictions about the structure of the rest of the organism from whence it came. We very properly ascribe this to natural selection. But it is worth noting that to a limited extent this same feature is true of even comparatively simple physical steady states; though the more unlike a system is to the human body the more difficult it is to perceive adaptation.

Let me illustrate this by a remarkable system which I am currently investigating. There is on the north coast of Norfolk a pond in fields behind sandhills which contains a relict marine fauna. Except for the brief great flood of the North Sea surge of 1953 it has had no contact with the sea. Old maps from Elizabethan times onwards suggest that this pond has been in existence for roughly 250 years. I am not concerned with the extremely interesting fauna and flora, but with the physical conditions of the pond itself which enable an isolated marine fauna to survive. The pond has a remarkable constancy of sea-water supply and of composition from springs, and an even more remarkable constancy of temperature, and a constancy of alkalinity. These constancies are ultimately due to the conformation of the coast, and particularly to the fact that the very extensive sands on the sea shore become saturated with sea water at the highest tides and slowly supply the pond at near high-water neap-tide level. The slow deep passage of the sea water through the sand almost certainly accounts for the constancy of temperature. The result of all this is that accidental circumstance has given rise to an almost ideal 'natural aquarium' for the preservation of marine organisms; one which would have cost great sums of money to build artificially; and one which contains a 'built-in' mechanism for the maintenance of constancy of conditions. Yet just such 'homoeostasis' mechanisms are one of the most important features of the tissues of living organisms, where their existence is ascribed to the adaptive effects of natural selection.

This is but a particular instance of a very general set of phenomena. Their interest lies in the fact that steady states in the physical environment can develop seeming 'adaptive' properties where neither the simple law of competition nor the effect of selection on reproducing heritable systems can operate. I have pointed out elsewhere[4] that such self-preserving adaptive features might sometimes appear with the mere lapse of time. If a configuration of matter and energy changes with time and in due course happens to reach a state which corresponds to a stage in which deviation leads to correcting change towards a mean, or if it happens to lead to a stage in a relaxation oscillation, the result will be an enduring system which we can observe as an object. And this will be true if during its history a dynamic material system in passing through successive configurations happens to reach one which by chance contains functionally stabilising mechanisms. Clearly this rule has something in common with the principle of natural selection, but it is by no means the same.

Other physical systems of this sort resemble living organisms in yet more features. Thunderstorms have individuality and functional parts. Under the conditions which engender them individual systems come locally into competition with each other for 'food supply', and some survive at the expense of others. But they are formed by spontaneous generation, and without the power of reproduction competitive survival depends upon terrestrial and aerial accidents of geography.

Despite their points of resemblance there is an enormous gap between such systems and living organisms. That difference cannot be adequately discussed here. But if, as in living organisms, systems are capable of reproduction and heritable variation—properties which, notwithstanding Lewes, present studies of the structure and chemical properties of nucleic acids are enabling us to understand—then natural selection will operate and evolutionary adaptation will proceed. I only wish to point out that the principle of natural selection is one of a related series of principles governing complex systems in the material world. In so far as these principles place restrictions upon the fate of material configurations of matter and energy they resemble and are additional to the laws of thermodynamics. Indeed these principles, particularly that of natural selection, have significant points in common, particularly with the second law. Both natural selection and the second law tell us the direction in which change takes place. A cinema film of the events depending on either is understandable if run forwards in time. It is magical and absurd if run backwards. Events which simply involve some other laws—such as the principle of conservation of energy—are reversible and do not have this directional quality. But though both the second law and the law of natural selection tell us the direction of change, they do not specify what the change will be.

I have tried to show that as we see it the material universe is characterised by many remarkable features. Structures show various grades of organisation and these display emergent properties: at least empirically we have to make new assumptions to describe them in addition to those needed to describe the simpler units of which they are composed. Similarly, to describe their fate in time we require certain empirical laws, like those of thermodynamics. These should include natural selection and other principles to cover the fate of matter and energy in the most complex systems. Each of these various laws involves assumptions that are independent of the others. We could conceive of a universe in which one or other of these laws would not obtain.

The apparent uniqueness of the universe primarily depends upon the fact that we can conceive of so many alternatives to it. We also note that repeatedly special features, which we could conceive to be otherwise, make it suitable and possible for living things. How far is this impression of uniqueness, and the linkage of that uniqueness with the necessary conditions for the existence of life, justified?

When we consider in what way the conditions of existence necessary for life are unique, there are two matters to be considered: the necessary conditions which make the continued existence of a living organism possible; and the necessary conditions by which a system with the attributes of a living organism can be achieved.

Living machinery is unquestionably complex; that is, its existence depends on the interaction of many functionally adapted parts. The possibility that without some directive principle, such, for instance, as natural selection, unorganised matter will take up a configuration of this class seems indefinitely remote, for experience leads us to conceive of innumerable alternative configurations outside this class. The same is, of course, true in lower degree of a machine such as a watch, in which the directive principle is supplied by an artificer.

To give concrete illustration to this uniqueness we may consider the material basis of behaviour. The higher animals possess eyes and other complex sense organs which collect information. The engineering specifications to which such instruments must conform are rigidly limited; yet the structure meets those specifications. It will be noted that the structure of any such instrument requires parts in permanent spatial relationship, such as can be given by solids or liquid interfaces alone.

The information from such instruments is employed in a brain which, whatever else it does, performs the task of a computer and predictor machine. From this aspect its function is to foretell the future with high probability, that is, to prophesy, basing prediction upon information from past experience and upon the assumption that the universe will continue to operate in the future as it has in the past. Moreover, several times and independently even in wasps, snails and other 'simple' creatures we see the transition of the basis of this prediction from response to simple stimuli to responses to objects in a 'world model', essentially similar to that with which our own common sense endows the world around us.[8]

Irrespective of whether a predictor machine is built of mechanical parts, of electronic circuits, or of nerve cells, certain essential patterns of configuration are necessary. Very different materials may be used to establish these; all must conform to one or another of a very limited set of specifications. The probability that unorganised matter will by chance meet such a specification seems again indefinitely remote, unless some directive principle allows the selection of configurations having these unique properties.

It is important to realise that such uniqueness attaches to the whole class of such configurations. In this universe it might be possible that several alternative sorts of behaving organism were possible. But the uniqueness of the class would remain. There is no need here to suppose that behaving organisms should be confined to the kind with which we are familiar.

For the construction of a computer, material structures are required which are able to store information. At any moment such structures must bear traces of their past history. That history necessarily confers individuality on the structure. For that reason individual atoms and particles of a lower order cannot satisfy the requirement; for these cannot be impressed with individual distinguishing marks. It is true that an atom may pass into higher or lower states of energy by the gain or loss of quanta. But the occurrence of this passage in a single atom is indeterminate and unpredictable, and therefore cannot be used to mark some other event. Whatever its state of energy, a single atom cannot inform us of its past history. In contrast, grosser multimolecular structures can retain manifold traces of their past in differences of configuration—as in a house that has been destroyed by fire.

A purely gaseous system, however, cannot by itself retain the necessary fixed relationship between its parts. Notwithstanding the complexity of some gaseous systems, as in thunderstorms, one cannot build a structure analogous to a slide rule or to an abacus from gases alone. For whatever part fluids may or may not play in the construction of a computing machine, solids which retain configurations are pre-eminently a necessity.

This limitation has an important consequence in that, viewed on the entire scale of temperature, there is an upper limit beyond which the solid, or even the liquid, state cannot be maintained. Indeed, solids may be said to retain some of that fixity of atomic relationship characteristic of substances at absolute zero. On the temperature scales of the universe computing machines are only possible near absolute zero.

Neither sensory instruments nor computing machinery consist simply of static configurations of solids. To operate, they must continually receive impressed changes. For this, energy is required and this in turn requires freedom of movement not to be found in static solid systems. The necessity for solids and the necessity for freedom of movement, which is to be found in fluids, set fairly narrow temperature limits within which sensory instruments and computing machinery can exist and operate, irrespective of whether we are concerned with living machinery as we know it or not.

Except in the sense that were it otherwise we, as observers, would not be present, there is no obvious reason why the universe should be such that such machinery can exist.

In living organisms as we know them, these rigorous requirements are met in systems which include not only solids, but also liquids, particularly water, with elaborate and functionally adapted interfaces. The entire living system is maintained as a steady state. For this, water has unique properties. In particular, its high dielectric constant allows oxidation and other energy-releasing processes which otherwise would only take place at a far higher temperature, at which the machinery would no longer be stable.

Liquid ammonia also has a high dielectric constant (though

much lower than that of water), and so for that matter does HCN. Conceivably liquid ammonia such as may perhaps exist on Jupiter could provide a medium for the existence of self-reproducing steady states. But to meet the requirements of complex future-directed behaviour as in animal life, there are further restrictions. Predictor machinery is not only primarily required for the acquisition of food, but there would be no profit in this unless the material acquired could be used as a source of energy. On earth this is rendered possible by the presence of oxygen—an atmospheric condition apparently unique at least in the planets of the solar system, and scarcely to be expected in a universe consisting predominantly of hydrogen. Even anaerobic bacteria are indirectly dependent upon the degradation of high-energy compounds resulting ultimately from a photosynthetic terrestrial system.

The actual structural elements of living organisms are based upon carbon compounds. The organic chemistry of carbon is uniquely suitable for the construction of complex systems. That property is shared by isotopes ^{12}C and ^{14}C, but the uniqueness resides in the electronic configuration common to these. Silicon is chemically not a substitute for it, and no other element provides such a necessary and unique collection of properties. Our present knowledge of enzyme action and of nucleic acids fully confirms the uniqueness of the properties of carbon.

Thus the conditions of existence in our universe have unique properties which render possible the existence of behaving living systems. We can say this even if there should prove to be more than one kind of 'living system' possible in the universe. On the other hand, the properties of water and carbon and oxygen provide unique properties which are necessary for the maintenance of living behaving systems. Advances in biochemistry have only increased the likelihood that if life has arisen elsewhere in the universe it is likely to share common features with that which we know: it would be likely to be a water–carbon compound system, and in the light of present knowledge there seem likely to be few if any alternatives for the evolution of adaptive organisms to that based upon nucleic acids; the special properties of these rest too firmly on the unique geometrical structure and electronic properties of certain few elements of the periodic table for us to see alternatives.

We have so far only considered some of the limitations imposed by the conditions of existence for the maintenance of a living system. Let us now consider how such a system could be evolved. Here again we can find severe general limitations to which any sort of reproducing adaptive system must conform, whether the particular system be like the organisms we know or not.

Natural selection provides a natural means of giving reality to possible adaptive structure. The basis of action depends upon heredity and variation. These requirements are independent of life being of the kind we know. Again, even in the simplest organisms there is advantage in the equivalent of some sexual exchange, because of the increased variation thereby offered to selection and the stability it confers on species. In terrestrial organisms that is achieved by exchange of nucleic acid; but again the advantage is independent of life being as we know it.

For selection to act there must be successive generations and the qualities of the species and of its individual varieties must be coded, so that by a sequence of events the coding of the initial configuration can lead to the adult individual. Codes of this sort are possible in gross matter, as in the templates and moulds from which machine parts are reproduced. These, however, suffer from the immense disadvantage that, like all gross enduring objects, they undergo denudation.

Construction of a heritable code capable of selective adaptation is impossible within the atom because, apart from the fact that there is no known way in which it can reproduce its like, it is not possible to impress new features upon its individuality. There is one level of organisation alone at which these two difficulties can be overcome; that is the molecular. By virtue of its atomic structure, a particular molecule can be repeated precisely, and it can be destroyed, but not denuded. On the other hand, complex molecules, particularly chain molecules, can be built up of sequences of more or less similar parts. There thus arises the possibility of families of very closely related molecules with closely related properties. The individuality of the links in each variant confers the possibility of a code, although as a molecule it is free from gradual denudation. Further, in the nucleic acid family, such a coded molecule is found to be capable of reproduction by the construction of another code as a mould in negative image of the first. From this the original can be re-formed without wear and tear. That, moreover, allows the code to be dynamically maintained, for if a link be destroyed during metabolic activity it may be replaced by an absolutely similar link.

Nevertheless, links in the chain may also suffer lasting change by substitution of a new part, and heritable variation is thus possible. But such a coded molecule must not only have the property that it can replicate itself, it must also have the particular functional property that it can act as a mould for the formation of other substances which, produced in succession, can lead to the construction of an individual of a specific organism. The sequences of the parts of the molecule confer uniqueness and the particular sequences have the power not only of reproduction but of ordered construction of a new individual.

Since the conceivable alternative sequences are extremely numerous, detailed resemblance between two nucleic acid molecules establishes a true homology at the molecular level of structure. By ensuring that a specific organism is the ultimate result of the operation of a particular nucleic acid chain, that chain exhibits adaptation in the Darwinian sense and is open to natural selection. All the same, none of this remarkable molecular machinery would be possible were it not for the unique properties of carbon, the unique energy exchanges possible with certain phosphates, and the unique ability of certain sequences of purines to allow replication and to synthesise other substances.

Irrespective of whether it were constructed of nucleic acid or not, any system with such properties could be the basis for species of 'living' organisms which selection could drive by adaptation under certain environmental conditions to achieve high emergent properties. In life as we know it, the conditions are in fact most remarkably fulfilled by the properties of the nucleic acid molecule. The manner in which this can give a heritable code, the power of replication, the initiation of a sequence of events leading to an adult organism, the power of variation, and hence the power of adaptation, offers an extraordinary collection of unique features, the absence of any one of which would render such a system impossible. Nor is it possible to propose any other element that can replace the unique fitness of carbon.

The recognition of uniqueness of this sort and the notion that such special features exhibited by the material world were susceptible of explanation is very old. Ray, Paley and the other natural theologians offered the particular explanation of a designer. Prominent in their argument was the analogy between the adapted structures of the animal body and the designed parts of machines. Darwin's theory of evolution recognised the significance of anatomical characteristics such as the unity of type and the fact of adaptation, but gave a different explanation congruent with a wealth of other evidence: natural selection.

But the operation of natural selection did not account for all the special features of the natural world. Fifty years ago L. J. Henderson raised again the problem of 'the fitness of the environment', using the term 'environment' as equivalent to the whole 'conditions of existence' which we have discussed. The old natural theologians had noted the remarkable collection of environmental properties that make the earth fit for the support of living things. Henderson points to the uniqueness of carbon; to the properties of water, not only as a medium but in such features as the relative lightness of ice with the consequent possibilities of oceanic circulation; and to the vast number of other special features. Since Henderson's time new unique features of the conditions of existence have been continually discovered; we may particularly note those which prescribe the biochemical basis of genetics.

These conditions of existence in this universe are not only unique, but this uniqueness is linked with the possibility of the existence of life and the evolutionary development of higher functions. We might have supposed that in the history of organisms some point might be reached where no further emergent properties in available configurations were possible to permit advance. Yet if in Silurian times some imaginary biologist's examination of the then existing cephalopods and annelids had led him to predict continued progress in behaviour machinery with new emergent features, he would not have been justified but he would have been proved right in the event.

It was with such things in mind that some years ago I argued as follows:[9]

The organism is thus built up of standard parts with unique properties. The older conceptions of evolutionary morphology stressed the graded adaptation of which the organism is capable, just as putty can be moulded to any desired shape. But the matters we have discussed lead us rather to consider the organism as more like a model made from a child's engineering constructional set: a set consisting of standard parts with unique properties, of strips, plates and wheels which can be utilised for various functional objectives, such as cranes and locomotives. Models made from such a set can in certain respects show graded adaptability, when the form of the model depends on a statistically large number of parts. But they also show certain severe limitations dependent on the restricted properties of the standard parts of the set.

Moreover, in this universe of ours any functional problem must be met by one or other of a few possible kinds of solution. If we want a bridge, it must be a suspension bridge, or a cantilever bridge, and so on. And the engineer who constructs the bridge must choose whichever of the solutions he can best employ with the standard parts at his disposal. In the design of a bridge, there are in fact three elements: the classes possible in this universe, the unique properties of the materials available for its construction; and the engineer only takes third place by selecting the class of solution, and by utilising the properties of his materials to achieve the job in hand. He is in a sense merely executing one of a set of blueprints already in abstract existence, though it requires insight to see that the blueprint is there.

We can apply these ideas to the construction of the living organism. Like all material structures they must conform to certain constructional principles. The standard parts available for the construction of organisms are the units of matter and energy which can exist only in certain possible configurations. Like the engineer, natural selection takes third place by giving reality to one or other of a series of possible structural solutions with the materials available. But the fact remains that we have arrived back at the eighteenth-century conception of an ideal plan as an essential constituent of organic design. Yet, as L. J. Henderson implies, any such plan is not peculiar to living things; it concerns the whole inanimate and animate universe.

In the child's constructional model of a crane we discern not one principle of design but at least two. For there is the design of the set of parts so that they shall build such things as cranes as well as the special design of the crane. In the living organism we can ascribe the apparent design of its immediately adaptive features to natural selection. Can we discern design in the properties of the units which make such an organism possible? These properties of the units are not the result of selection in the Darwinian sense. And if we see design in them we must say with Du Bois-Reymond: '...whoever gives only his little finger to teleology, will inevitably arrive at Paley's discarded "natural theology"...' Natural selection bowed Paley and his argument from design out of the front door in 1859; and here he would come climbing in through the back window saying that he owns the title deeds of the whole estate! Fortunately it is for the metaphysicians to examine his claim and not for me...

Clearly the course of evolutionary adaptation depends primarily upon this remarkable plan of possible configurations and on the restricted possible ways of meeting the engineering specification. It is by no means completely described by the principle of natural selection alone.

Henderson had concluded that the physical environment and the conditions of existence seem to offer a very large number of unique features fit for the existence of life. He also pointed out that whereas the unique adaptations of organisms can be attributed to natural selection, with the implication of a vast number of alternatives from which the fittest are selected, he found no evidence of this in the physical world; rather we find few alternatives which happen, remarkably, to include those suitable for the development of complex organisms. There is no obvious cause for linkage between the arbitrary properties of material configurations and the necessary conditions for living systems which selection must provide to give special adaptation of a particular species.

This same lack of a necessary cause for linkage seems also to be found in the simpler, though complex, physical systems I have discussed.

It is important to say that in this argument neither Henderson nor I am necessarily inferring the existence of a designer. What we both point out is that in our present state of knowledge both organism and environment show an abundance of unique necessary properties for life; and that natural selection alone does not account for these. In parenthesis I would say that Dr Lack, in his recent essay *Evolutionary Theory and Christian Belief*, is mistaken when he says that we draw an inevitable inference of the existence of a designer as the necessary consequence of the situation we try to disclose. Nor do we assume, as he suggests we do, that there can be no life except in forms that we know—the possibility of self-reproducing systems being generated in another medium of high dielectric constant, say, liquid ammonia, on the planet Jupiter is conceivable. The special conditions which fit our universe for the existence of life and of conscious beings include many which are more fundamental than those needed for the particular existence of life as we know it.

The unique element carbon in the periodic table and the substances derived from it alone appear to satisfy the requirement of production of some such coding molecule as nucleic acid. And even if there prove to be one or two entirely alternative ways of constructing self-reproducing systems in our universe, we should still be faced with the uniqueness of the emergent properties of configuration and the laws of thermodynamics; and there is no obvious reason to account for their evident linkage with the possibility of living systems. There are so many ways in which self-reproducing, selectable and therefore adaptable systems could be rendered impossible if we suppose one or other empirically observed fact about the universe to be not so.

The different grades of system, atoms, molecules, organisms and so on, display empirically emergent properties, so that fresh assumptions have to be made in hypotheses representing them over and above those needed for the lower-order systems from which they are built. When we examine these systems we find a linkage of the properties of these systems with those needed for the maintenance of life and the more complex phenomena of living creatures. If we were to suppose a universe in which at least certain of the properties presented to us by space, matter and energy were not so, material self-reproducing systems capable of evolving would seem impossible. It is natural to ask why such linkage seems to exist even if in the end no answer can be given, or even if such a question proves unreal and a consequence of our ignorance.

One set of phenomena which calls for explanation, the mutual adaptation of the parts of animals for function, follows indeed from natural selection, provided the universe is such that the right materials are available and appropriate engineering solutions are possible. But this 'fitness' of the conditions of existence includes remarkable features, as in the emergence of individual consciousness; and the unique properties of carbon, hydrogen and oxygen still seem to require explanation.

All we have is a specification that any explanation must meet: that the properties of the material universe are uniquely suitable for the evolution of living creatures. To be of scientific value any explanation must have predictable consequences. These do not seem to be attainable. If we could know that our own universe was only one of an indefinite number with varying properties we could perhaps invoke a solution analogous to the principle of natural selection: that only in certain universes, which happen to include ours, are the conditions suitable for the existence of life, and unless those condition are fulfilled there will be no observers to note the fact. But even if there were any conceivable way of testing such an hypothesis we should only have put off the problem of why, in all those universes, our own should be possible.

Another suggestion—for there may be yet others, unknown—is the pre-Darwinian one based on the analogy of the unique suitableness of material systems for life and the unique suitableness of the parts of a machine constructed by an artificer. In the case of the human artificer, his design is based upon past acquaintance with the material world, including the appearance of special properties in ordered structures, and the possible existence of disordered structures as well. It is also based upon a grand assurance that our universe and all its properties will continue as it is at present and will not abruptly change its character. Scientifically, there seems no way of testing the assumption that the fitness of the conditions of existence is the result of intelligent design. Equally, there seems no way of proving that this is not so.

But lest we dismiss the problem of the fitness of conditions for life itself and for the emergence of the human faculty, it must be remembered that whatever we can or cannot do we cannot prove the problem to be inscrutable. Nor must it be supposed that every generation forever comes back to the position of its predecessors without progress. Our view of the nature of life itself differs from that of our grandfathers not merely in the accretion of fact but in the quality of our understanding of it. Further, any analysis such as I have attempted necessarily suffers limitations. It is not only that in these specialist days it is perhaps an impertinence for a biologist to attempt this sort of thing. It is that, just as Thomas Henry Huxley once reminded us that we cannot get more out of the mathematical mill than we put into it, so that same limitation applies to all our hypotheses. Good and evil, beauty and ugliness, intention and responsibility are matters of direct experience. If our hypotheses fail to represent them as we are aware of them in our lives, it is too bad—for our hypotheses.

Perhaps a way to advance may be by examination of the moral features of the scientific method. We attain knowledge of the natural world by simultaneously employing two distinct methods. There is that which is particularly evident in the naturalist: the qualitative perception of relationships. This has the virtue that it respects the phenomena we witness as well as our ideas about them. It has the danger that the value we place on the facts is for us to decide. Integrity is essential; without it we shall only find what we intend. The other method is that of construction of hypotheses by logical development from accepted assumptions. The present mathematical models of the physicist are grand examples of this. It has the virtue that we are compelled to face the logical consequences of our assumptions. It has the danger that we may tend to prefer to study only those aspects of nature which yield hypotheses that relieve us from the necessity of decision, rather than those which most closely represent the whole of nature. Every man of science knows the feeling of disappointment when his simple and elegant hypothesis proves inadequate in a busy world. There is also the danger that by this second method alone we systematically attempt only the easy question in the natural examination paper. It is so much easier to advance our knowledge of the atomic nucleus than to predict the ecological consequences of what we do with the knowledge.

APPENDIX 2

Learning, world-models and pre-adaptation

TO THE GREAT pioneer students of the behaviour of the lower animals, Jennings, Parker, Romanes and Charles Darwin himself, the programme of this symposium would have been exciting. It is not only the success in recent years of methods of analysis which they themselves foreshadowed, nor even the unexpected information which new methods of analysis have given us about the nature of learning and of the higher attributes which we notice in animal behaviour. It is also that in the behaviour of the lower organisms we begin to see the variety of means by which these attributes are derived.

Concerned, as we primarily are, with the origin and nature of the human faculty, it is salutary to consider that certain of the features which we note as characteristic of it are not even dependent on the presence of a nervous system. As Charles Darwin[1] said of the directed activities associated with food-capture in the sundew *Drosera rotundifolia*, 'This may be called reflex action, though probably very different from that proceeding from the nerve-ganglion of an animal'. Further, however complex the behaviour machinery of the highest organisms may be, it follows from the principle of natural selection that its elaboration during the course of evolution could only have been possible through the selection of pre-existing properties in the functional systems of the organism; properties which did not themselves endow organisms at that stage with machinery for the higher attributes of behaviour, but which through chance variation and selection would provide a basis from which they could be directly developed. Such properties, the so-called 'pre-adaptations' of Cuénot,[2] are to be sought particularly in the simpler behaving organisms. It seems useful, therefore, to reflect not only upon the phenomena of learning in various organisms, but also upon the varied means by which they achieve such qualities, and the properties offered by the organisation of simpler organisms which under the influence of variation and selection could lead to signal advances in behaviour machinery in more highly developed organisms.

Whatever its basis, learning is characteristic of the animal 'habit'. Plants, and the generality of micro-organisms, are essentially molecular machines, the manifest adaptive organisation of which is concerned with events at a molecular level, as in the carbon dioxide absorption and photosynthetic machinery of the leaves of plants. In contrast, animals are essentially predatory behaviour machines, the manifest adaptive organisation of which is of a grosser order of size, as in the sense organs, brain and raptorial organs of our own bodies. The primary need of a predatory behaviour machine is to foretell the future, where food is and where harm is not and so on. Such prophecy is possible with high probability on assumptions based on experience: (1) Information collected about past events can be a guide to the course of events in the future; the rules which seem so far to have governed the universe will not abruptly change and make nonsense of past experience as a guide. (2) However different another organism is from ourselves and however different the sources of information which it appears to utilise, on analysis its behaviour is completely consistent with the occurrence of the external objects and events presented to us by the naïve realism implicit in our own everyday behaviour. However private is the world of an ant, of *Paramoecium* and of another man, individual stones, puddles, plants and men exist for them all: these essential features of the external world are independent of the nature of the organism.

We can go further than this. Analysis has always shown that external objects, including other organisms and also the other parts of our own bodies, are composed of the same classes of material and subject to the same physical laws and restrictions.[3] Thus, specification for the behaviour machinery set by natural selection requires the organism to use the same classes of raptorial machinery to obtain the same sources of matter and energy for its maintenance and by employing the same possible classes of available predictor mechanism, basing that prediction on the same fund of information available in the present and past configurations of matter and energy.

The future-directed character of an organism is not only concerned with behaviour. It extends to its structure and life history, as in the marvellous adaptations of a parasite. We may interpret this as the organisation of structure and life history for future ends through the operation of natural selection on the species. But the activity or behaviour of the organism is more strikingly future-directed. This may happen in more than one way. There may be reflexes, drives or instincts which are inborn movement-forms, the 'fixed action patterns' of Lorenz, in response to stimuli or releasers of varying grades of complexity. But the development of such a response appropriate to a given premonitory stimulus is not in this case dependent on the experience of the individuals; it is the result of enforced adaptation in the species. We may interpret it as the organisation of behaviour as the result of natural selection on its machinery, within the species.

The second way in which behaviour may become future-directed is through 'learning'. Many definitions of this have been given; almost all, as Thorpe[4] points out, are unsatisfactory. His own definition is in fact excellent: that we may interpret learned behaviour as 'the organisation of behaviour as the result of individual experience'. Thorpe points out that such a definition may seem inadaquate. It seems to be too all-embracing and to miss certain undefined characteristic features of learning. But the fault is not in the definition, it is in our inclusion within a single category of an enormous range of kind and of complexity of future-directed behaviour influenced by individual experience. For any organism, not only the capacity but also the quality of its learning is a function of the experience possible to that organism and to the engineering possibilities inherent in its grade of structure. Thorpe's definition is excellent because it gives the minimum qualities needed for a phenomenon to be classed as an example of learning, and it does not fall into the common error of entangling a definition with contemporary hypotheses of how the phenomenon is brought about. It is in fact a waste of time to try to elaborate this definition.

The importance of this becomes evident as soon as we study learning in the simplest animals. An astonishingly wide variety of organisms are said to show evidence of the power to learn from past experience, including unicellular organisms, which possess no nervous system. Dr Jensen points out in this symposium that supposed examples of learning in Protozoa repeatedly turn out to be based on inadequate controls. But there are certain examples which are not so easily dismissed. Jennings[5] in his classical work on *Stentor* showed that repeated noxious stimulation by carmine, or by other means, led to a succession of varied activities directed towards the evasion of the stimulus:

> The same individual does not always behave in the same way under the same external conditions, but the behavior depends upon the physiological condition of the animal. The reaction to any given stimulus is modified by the past experience of the animal, and the modifications are regulatory, not haphazard, in character. The phenomena are thus similar to those shown in the 'learning' of higher organisms, save that the modifications depend upon less complex relations and last a shorter time.

Such phenomena undoubtedly fall within Thorpe's definition of learning. A second case concerns the progressive increase in speed with which a *Paramoecium* confined within a capillary tube manages to execute turning movements with successive trials. Buytendijk[6] concluded that this resulted from increased flexibility of the organism consequent on mechanical stimulation, and could be a purely physical effect. Nevertheless, the effect would seem to be of adaptive value for escape and formally it falls within Thorpe's definition. If we say that the effect is of too simple a physical origin to be 'true learning' we are in danger of entangling our definition with such provisional models as we have in mind of the physical basis of learning—the very matter we seek to investigate.

Whatever the nature of the learning process, it is operated by the physical structure of the organism. Our job is to understand that machinery, to see our way to the making of a mechanical model of it. Lord Kelvin, the great physicist of the last century, once said: 'It seems to me that the test of, "Do we understand a particular point in physics?" is, "Can we make a mechanical model of it?".'

Today the physicist prefers a purely mathematical model, pointing out the danger of unconsciously endowing phenomena with irrelevant properties arising from mechanical models; the billiard balls that represent molecules must not carry their ivory or their colour into our arguments. All the same, one has but to examine models of the cell membrane and nervous models of the learning process to realise that the biologist repeatedly uses mechanical models. For these have certain great advantages. First, a mechanical model which can be seen to work provides a guarantee that we have not unwittingly made false assumptions which, at least for the ordinary man of science, are more easily made with an abstract model. Secondly, the mechanical model focuses our attention on the fact that our main task is to determine the physical class to which a phenomenon belongs. The foundation of our understanding of nervous action rests on the demonstration by Keith Lucas and Adrian[7] that the nervous impulse belonged to the class of a conducted disturbance—such as one sees traversing an ignited gunpowder train.

But this determination of the class to which a phenomenon belongs is much more difficult in biology than in physical science. We have no guarantee that the machinery of learning, the adaptive modification of behaviour through past experience, always belongs to a single class.[8] It is of selective advantage to be able to learn. Natural selection is not concerned with how this requirement is met. Different animals may meet the imposed specification in different ways; and different means may be used to meet it in different reactions even in the same organism. A phenomenon such as habituation may result from anything from sensory accommodation up to a highly complex central nervous effect. The difference between these is not in the habituation effect, but in the complexity and subtlety of the stimulus recognised as non-significant by the animal.

Again, we repeatedly find that advantageous physiological activities can be effected by an organism in more than one way. In a simple physical system causes may seem simple: remove the sodium and the lines at 5890 and 5896 Å disappear from the flame. In a living system there are often alternative routes, so that if one method of achieving the goal is stopped another alternative stands ready to take its place. That is true even of oxidation processes in the cell. Further, it not infrequently happens that the simultaneous presence of two alternative kinds of machine has greater value than either by itself. I have pointed out elsewhere[9] that in the origin of the nervous system we do not find direct muscular activity superseded by sensory-muscular nervous arcs: the simultaneous presence of both in the one system gives these simpler Metazoa much greater flexibility of behaviour.

The biologist certainly uses models, whether his attack is essentially by describing phenomena and then drawing analogies, as the hydraulic analogies that have been used to illustrate 'drive' and other behaviour phenomena, or by first deriving his model from formal assumptions needed for the construction of predictor machinery and then seeing how far this describes the features of central nervous action. But in all cases, the complication of the living machinery for which a model is being made places the biologist at a great disadvantage compared with the physicist who confines himself to simple systems. The complication arises from the operation of natural selection. The system selected is an 'improbable' configuration compared with a simple physical system because conceivable configurations of physical systems which have the directive qualities of a machine are necessarily far fewer than those which have not. This not only gives rise to the difficulties already mentioned, but makes the scientific method itself more difficult of application.

Following the principle of economy of hypothesis we choose what seems the simplest explanation for a phenomenon. We go further, and choose the simplest phenomena for our first analysis.

Thereby we may systematically oversimplify our interpretation.[10] Thus, in the analysis of the behaviour of an anthozoan I began by attacking the simple reflex closure. A comparatively simple system was disclosed. But this did not show that anthozoan behaviour was based on a system of simple reflexes, as might at first be supposed. It was subsequently found that spontaneous rhythmic activity played a large part in the behaviour of these animals,[11] and recently Passano and McCullough[12] have given reason to suppose that in *Hydra* a variety of rhythmic electrical potentials are an essential part of its various activities.

Similarly, a hypothetical neurological model for a conditioned reflex is easy enough to invent so long as the conditioned and unconditioned stimuli are simple, as in an electric shock. But as soon as complex stimuli are considered, the implications of widely diffused neurological contacts make it far harder to envisage a satisfactory model. Our difficulties become acute when we consider the implication of such observations as those of Thorpe[13] on the apparent knowledge of locality which determines behaviour in the hunting wasp *Ammophila pubescens*. This animal behaves as though it was possessed of a map or model of the region round its nest. The same is true of the digger wasp *Bembix rostrata* discussed by Dr van Iersel in this symposium. His work throws light upon the manner in which the map is coded. In a sense, these are of course examples of a learning process. But their really interesting feature is not the mere retention of past information, but the model of the external world that the insect has abstracted from it.

Is our first step, in fact, to try and make a model or models of the learning processes? Nothing is easier than to provide mechanical models for storage of information. The magnetic tape, the silica-mercury acoustic devices used in computing machines, photographs and many other devices can store information. How far any of these have any features analogous with those of information storage in animals for the moment does not matter. For the animal there are neurological models for storing information, such as patterns of synaptic resistance, the self-exciting neuronal circuits of Lorente de Nó,[14] the oscillating circuits of Pringle,[15] and so on. At this conference is discussed a new possible store, RNA, though doubt about the validity of the experiments on which the conclusions are based is not yet removed.

These models provide possible ways of storing information in the nervous system. Comparison with mechanical models may help us to understand the classes to which parts of the behaviour machinery belong. But our main problem, even in simple animals, is not simply: How does the animal store information? It is: What is the nature of the information stored and what does the nervous system do with it?

What in fact is the behaviour machine of which we wish to make a model? As I have said, it is primarily a predictor machine telling the organism what will happen, and doing so on the basis of present patterns of information, together with past sequences of patterns from which rules governing the present sequence may be inferred. Craik,[16] in his invaluable analysis of the nature of human thought, points out the implication that this involves a model of the external world; a model derived from the patterns we have just discussed, one which is not static but which can be used to predict the future. Thereby we can 'try out' consequences which have not yet occurred. In operation the model in fact allows some degree of extension into the future. It makes use of memory, but memory which is not purely static records of particular past configurations, simple sequences in time, but the sequences of many dimensions in time, from whence rules governing the probable future can be inferred.

That the essential machinery we are trying to elucidate is a model of the world around the organism with the organism in it seems to some extent true of the behaviour of many organisms. The behaviour of *Ammophila* and *Bembix* seems to require just that. Memory and learning are important in this model, but they are not the whole model.

Important as is the problem of finding various models to represent the different kinds of learning process so that we can determine to what physical classes these phenomena belong, I suggest that we may get further if we ask ourselves at what point in the scale of complexity of neurological organisation in animals we are forced to assume that an animal's behaviour implies the possession of an internal model representing the current events in space and time. At present we have no evidence that the assumption of such a model is necessary to describe the future-directed behaviour of a protozoan or of a coelenterate. But the assumption of such a model does seem necessary in the behaviour of *Ammophila*. To imagine a physical model which would possess such properties is certainly difficult. On the other hand the quantity of nervous elements involved in the behaviour machinery of insects is surprisingly small. Charles Darwin himself remarked:[17]

It is certain that there may be extraordinary mental activity with an extremely small absolute mass of nervous matter: thus the wonderfully diversified instincts, mental powers, and affections of ants are notorious, yet their cerebral ganglia are not so large as the quarter of a small pin's head. Under this point of view, the brain of an ant is one of the most marvellous atoms of matter in the world, perhaps more so than the brain of man.

It is curious how little attention has been paid to this important fact. Our knowledge of the size and number of nerve cells in the brains of animals is exceedingly scanty and mostly rests on estimates of long ago. Even the common statement of ten thousand million as the number of nerve cells in the human brain rests on an estimate of over sixty years ago.[18] Young has recently given us an estimate of 5×10^6 nerve cells for the nervous system of *Octopus*.[19] But for insects the position is much less satisfactory. For the estimated size of insect brains we go back more than one hundred years to Dujardin,[20] who gives the volume of the brain of the honeybee, *Apis* sp., as 0·62 cu.mm. and that of the ant, *Formica* sp., as 0·065 cu.mm. The brain of man has a volume of about 1600 ml. Accepting Dujardin's figures, the brain of the bee is about 4×10^{-7} the volume of that of a man, and the brain of an ant is about 4×10^{-8} of that of man. Even though the size of a nerve-cell body is very far from being proportional to the size of the organism, it would seem impossible to be forced to conclude from the above figures that a bee-brain contains only a few thousand ganglion cells and that of an ant only a few hundred: the larger the brain the larger is the proportion which must be assigned to connecting axons and the smaller the number of cells per unit volume.

I have been unable to find data upon which to make a satisfactory estimate of the number of nerve-cell bodies in a bee's brain. Such an estimate is urgently needed in view of the enormous complexity of behaviour of these animals which has become apparent following the work of von Frisch. But the point is of such interest that even a rough guess may prove useful. Rockstein[21] gives figures for the number of cells in two arbitrarily placed 5μ transverse sections across the brain of *Apis*. The numbers obtained are about 800 cells per section; since he states that the smallest neurone cells have a nuclear diameter of 5μ, this number is likely to exceed the actual mean number of cells per 5μ section. But if we take a mean value of 800 cells per section, and take Dujardin's value for the size of the brain and allow for its shape, we arrive at a value of the order of 10^5 nerve cells. The number in the brain of *Formica* is likely to be very much smaller. Such very rough estimates point to the great importance of getting accurate ones. Of course the individual nerve cells are not simple standard units, the same in every organism and the same in all its nervous parts. The pattern of its axons and dendrites is of essential importance to the significance of a nerve cell as a unit. All the same, a configuration of 10^5 structures is by no means beyond our power of apprehension; and when we bear in mind that a really large computing machine such as 'Titan' is built up from units of the order of 10^6 in number we realise the significance of Darwin's comment, however different the units may be. Difficult as is the problem of finding a model which represents the bee's world in such a brain, such numbers make one feel that the principles of its machinery should be within our power to grasp. Indeed, when someone enables us so to do, we may well find ourselves repeating the comment of Thomas Henry Huxley after reading *The Origin of Species*: 'How extremely stupid not to have thought of that!'

There are two other points: study of our histological preparations would suggest that 10^5 nerve cells cannot vastly exceed the number of nerve cells in the nerve net of a sea-anemone. If we find that this model of the real world appears somewhere in grade between such animals and bees it does not seem that this new faculty is the result of a vast increase in the number of nerve cells, but rather of their pattern of organisation: secondly, bearing in mind what a bee can do with a brain of less than one cubic millimetre, we may properly ask ourselves, What are the remaining million or so cubic millimetres of our own brain doing?

Whether in *Ammophila* or man, the internal model of the real world is derived from changing patterns of sensory stimulation. Nothing can be more private to the organism and nothing could be more varied than these patterns of individual stimuli. In fact, as Adrian[22] noted, from the outset information conveyed by such stimuli undergoes analysis. Different rates of sensory adaptation give information of rate of change with respect to time as well as the bare fact of stimulation. The receptors of the mammalian retina include not only those that respond to 'on' illumination but also to 'off' and to 'on/off'. Such differentiation enables the central nervous system to record sequences of pattern, and if the effects of previous sequences have been stored, the system has the necessary information for the detection of recurrent relationships in the pattern. Eccles[23] has given a useful hypothesis of the manner in which, in a complex interlacing neuronal network, the same nerve cells can serve to record different sensory patterns, and the manner in which, through physical changes in the synaptic junctions, recurrent patterns may retain a record in the central nervous system.

Such abstraction of common and recurrent features in the initial pattern of stimuli would seem to have two important features. First, however different the varied successive patterns of stimuli may be, as I have said, experience leads us to suppose that they all arise from the same world around us with the same objects. Successive abstraction from different sets of stimuli provides, as it were, an 'envelope' defining the objects of what we call the 'real' world. The patterns of stimuli received do not disclose a series of arbitrary relationships, but a coherent one conforming to rules imposed by the coherent conditions of existence. The detection of these relationships is made the easier when, as in the retina, the same rules governing spatial relationships enable the construction of an instrument in which events closely related spatially are recorded in correspondingly closely spatially related receptors, with all the simplification of neuronal connections that makes possible. Thus the coherence and unity which experience teaches us to obtain in the world of external phenomena carry a coherence and unity into the model in the central nervous system.

Secondly, given the coherence and unity of the external world, the internal models in different organisms will approach each other more and more in kind the more completely the information of the sensory stimuli of each is abstracted. Finally, from the point of view of learning and behaviour, the higher the grade of the organism the more it will respond to abstracted 'things', detection of the character of which is the result of experience, rather than to mere initial stimuli. I instantly perceive a crab to be *Carcinus maenas* and do so in a way quite different from that which I used in childhood. Experience has in fact altered the machinery of perception.[24]

Though it seems clear that the behaviour of *Ammophila* and *Apis* implies some internal model of the real world and the character of its events, there is at present no obvious trace of such a model in the behaviour of a coelenterate or a protozoan, despite the fact that such simple creatures as *Stentor* show evidence of learning. At some point in the evolution of higher organisms this kind of transition has several times been made. It must have involved the utilisation of certain pre-adaptive qualities in the behaviour machinery of the simplest animals. What is needed is not merely an analysis of the power to learn, but more examination of the degree of abstraction from the pattern of direct stimuli to which the animal actually responds. To what degree of abstraction from the patterns of its sensory input does *Ammophila*, a starfish, a sea-anemone, or *Stentor* react? But certain phenomena in coelenterates are of considerable interest.

In the first place, the mammalian cerebral cortex shows resemblance to a nerve net both in its structure and in the spread of excitation in it.[23] Eccles suggests that perception and memory are to be related to patterns of excitation in this network, some of which, through structural changes in the synapses, can acquire some degree of permanence through repeated associations of particular kinds of stimulation pattern. Now, in the coelenterate nerve network, Horridge[25] by considering models of such networks has pointed out that, according to its junctional properties and the character and duration of the stimulus, different patterns of conduction are to be expected and that many of these are actually to be found. Josephson, Reiss and Worthy[26] have further noted that an even more satisfactory correspondence is obtained if full allowance is made for the property of 'interneural facilitation'.[27] Moreover, he notes that in such a model network, when conduction is limited, the spread of excitation may be highly asymmetrical in random directions, notwithstanding the uniform properties conferred on the network in the model. Whilst these phenomena in coelenterates are mainly concerned with motor responses, on the sensory side such temporary conduction paths connecting different regions of a network could provide a basis for the patterns of excitation envisaged by Eccles[23] in the mammalian cortex.

Apart from this, in neither the coelenterate nerve net nor the mammalian cortex do we in fact have a uniform structure. Conduction in coelenterates takes place in certain favoured directions related to the anatomical structure of the organism.[28] Nerve networks in coelenterates, which in these animals are directly connected with receptor cells and with the muscle sheets,[12] thus seem to have properties in common with the nerve network of the central nervous system in higher animals which controls complex behaviour.

There is a second feature of the coelenterate excitation system which resembles phenomena in central nervous systems. In the Mammalia, there occur rhythmic electrical potentials in the resting nerve network which constitutes the cortex of the brain. Sensory stimulation and purposeful activity tend to abolish the rhythm, which is seemingly replaced by irregular patterns of activity.[22,23] Whatever the explanation of these electrical phenomena, they are certainly concerned with that part of the nervous system associated with perception and complex behaviour. Now, although there is no clear evidence of learning in the coelenterates, their nerve nets do have certain features in common with this. Passano & McCullough[12] and others have shown that rhythmic electrical activity is a normal feature even of *Hydra*. Batham and Pantin[29] showed that different patterns of rhythmic activity play an important part in the behaviour of Anthozoans. Modification of essentially the same rhythmic patterns in the neuromuscular system can lead to such different activities as general tonic form, digestive peristalsis, seeking activities for food and locomotion; and Robson[30,31] has shown how complex rhythms of the same sort lead to swimming in *Stomphia*. None of these motor activities is connected with learning, but we have at least this much parallel with the central nervous system, that the behaviour machine involves rhythmic activities which can be modified in various ways. If we knew what the rhythms really signify in the higher kinds of nervous system, it might well be that these rhythms of the coelenterata, as well as the conduction phenomena of their networks, present pre-adaptive features out of which the central nervous activity of more complex organisms could be evolved.

SUMMARY

1. Animals are essentially predatory behaviour machines which make predictions necessary for future-directed behaviour on the basis that information collected about past events can be a guide to future events.

2. Future-directed behaviour, if not 'built-in' reflexes, etc., demands the organisation of behaviour as a result of individual experience. The machinery of this may vary greatly in different animals. Attempts to define learning more precisely often fall into the error of entangling the definition with contemporary hypotheses of how learning is brought about in particular cases.

The difficulties are particularly clear in the study of learning in Protozoa.

3. It is of selective advantage to learn. Natural selection is primarily concerned only that this requirement be met. It may be met by several classes of machine. There is no single 'mechanism of learning'.

4. Nevertheless, to analyse the physiological machinery of learning in any particular case the biologist uses models and analogies. This is much more difficult a matter than the provision of models for simple physical phenomena.

5. So far as learning depends on storage of information there are many models, physical and neurological, to choose from. But our main problem is not how an animal stores information, but what is the nature of the information stored and what does the nervous system do with it. Consideration of the behaviour of solitary wasps shows that these organisms seem to possess a map or model of the region round the nest. The important problem for us is the nature of that model in neurological terms, rather than how information is stored.

6. The surprising fact emerges that notwithstanding the complexity of the behaviour of ants, bees and wasps the size of the brain is minute. There are few even rough estimates of the number of nerve cells in the brains of invertebrates. Very rough estimates suggest about 10^5 nerve cells for the bee's brain and far fewer for small hymenoptera. Configurations of numbers of parts of this order should not be too hard for us to grasp, notwithstanding the special features of nerve cells, their dendrites and axons.

7. It is pointed out that the 'real world' in which a man finds himself is essentially the same as that in which a wasp or other organism finds itself. However different their sense organs and the sensory information which they gain, the 'real world of external objects' which the various nervous systems extract from that information is essentially the same.

8. Though it is clear that in man, and in the bee, their behaviour implies some internal model of this real world and the character of its events, there is at present no obvious trace of such a model

implicit in the behaviour of a coelenterate or protozoan. At what point in the evolution of various phyla of animals did such a model arise? What is needed is not merely analysis of the development of the power to learn, but a study of the extent to which different kinds of animal are able to abstract a model of the external world from the information they receive.

9. Although so far coelenterates show no clear evidence of power to abstract such a model from their sensory stimuli, or indeed of any power to learn, nevertheless the properties of their nerve nets and the recently discovered property of electrical rhythms in them show significant similarities to structural and functional features of the nervous organisation of the mammalian cortex. Possibly we have here in the coelenterata some of the 'pre-adaptive' features which enabled the central nervous system in more than one class of animal independently to reach complex behaviour.

APPENDIX 3

Organism and environment

CHARLES DARWIN entitled his work: *On the Origin of Species by Means of Natural Selection, or the Preservation of Favoured Races in the Struggle for Life*. In discussing Natural Selection, he says: 'Let it be borne in mind how infinitely close-fitting are the mutual relations of all organic beings to each other and to their physical conditions of life.'[1] Those 'close-fitting' relations were the result of natural selection. Long after, in 1913, L. J. Henderson, in that remarkable book *The Fitness of the Environment*, noted that, while biologists had given much thought to the adaptations of the living organism to the environment, they had given little to the nature of the environment itself.

But although Darwin's fitness involves that which fits and that which is fitted, or more correctly a reciprocal relationship, it has been the habit of biologists since Darwin to consider only the adaptations of the living organisms to the environment. For them, in fact, the environment, in its past, present, and future, has been an independent variable, and it has not entered into any of the modern speculations to consider if by chance the material universe also may be subjected to laws which are in the largest sense important in organic evolution. Yet fitness there must be, in environment as well as in the organism. How, for example, could man adapt his civilization to water power, if no water power existed within his reach?[2]

Surprisingly, this absence of attention still continues. In pre-Darwinian days it was not so, and the peculiar fitness of the environment for living things was as well recognised as was the apparent element of design in organisms themselves. But consideration of Henderson's statement shows us that there are two very different aspects of the environment which have not been properly separated. There is (1) the world of physical objects, including other creatures, by which an organism is surrounded, and (2) the set of conditions, peculiar to this universe, governing organism and external world alike. Irrespective of the fact that living organisms may display additional special features of their own, in both organism and environment, the same kinds of matter and energy appear to follow the same 'laws' and changes in time. Indeed, one of the most remarkable conclusions of astronomy is still that even in the most strange and distant galaxies we find the same elements and the same kinds of energy following the same familiar configurations.

Moreover, as I have said, there are only certain possible configurations with certain properties. The conditions of existence are such that there are only a limited number of real solutions to the engineering problems confronting the construction of a machine, or to the viable construction of a living organism. There are only certain ways in which an eye or camera can be made and there are only certain ways in which a computing machine designed to predict the future can be made. In living organisms such limitations determine the possible solutions to the engineering problems of an animal for it to be a successful predatory behaviour machine. It is convenient to speak of these as the conditions of existence to which both living and non-living matter are subject, and to distinguish this from the environment of external objects which surround an organism or a non-living entity.

There is a boundary, though not a precise one, between organism and environment. Even in the physical world the boundary of objects is not wholly precise. Michael Faraday noted long ago in his attack on Dalton's atomic theory that we only know an object by the forces it exerts and that these might extend, with attenuation, throughout the universe.[3]

In the prosecution of the physical sciences this external environment is commonly made as simple as possible by the observer, so that the number of necessary controls is few. For the biologist generally, the environment of his organisms is exceedingly complex—and he must put up with it. This may be overlooked in some analytical biological work. Thus in genetics the environment is at times treated as a simple constant thing providing an asymptote for natural selection. In fact it is exceedingly complex in both time and space: the same species may be found in environments which vary discontinuously. A successful species of bacterium may be found developing in very different food sources. The action of natural selection in the field is far more complex than the selective preservation of mutant *Drosophila* in an experiment.

In all this description, it will be noticed that we have tacitly assented to an external world of real objects. The attitude to this external world varies greatly in the different sciences, a fact of particular importance today, in which science itself and its nature are popularly identified with the technically successful parts of physical science. As I said in a recent essay:

> When we look at the different sciences a very important distinction becomes evident. Natural phenomena are extremely complex. The physical sciences as they now dominate us have achieved their rapid success in a great measure by deliberately restricting their attack to simple systems, thereby excluding many classes of natural phenomena from their study. Until, in the end, the nuclear physicist has to take into account the fact that the observer is a biological system, there is no need for him to burden his hypotheses with other sciences. There is no need for him to know any ecology or comparative anatomy. Because of this I speak of physics as one of the 'restricted sciences'. Biology and geology on the other hand are among the 'unrestricted sciences'. The solution of their problems may at any moment force biologists to study physics, chemistry, mathematics or any branch of human learning, just as Louis Pasteur had to become bacteriologist, entomologist, chemist, biochemist and physicist to achieve his goal. Almost every biological problem is a piece of operational research using other sciences for its solution.[4]

It should, of course, be borne in mind that certain sciences, such as meteorology, commonly classed in the physical sciences, are, in fact, unrestricted, involving as they do biology, geology, and so on. But their very complexity places them outside the pure physical sciences. Likewise, the development of certain special fields of biology may lead to these becoming to some degree restricted, as was at one period the case with taxonomy, and as is at present the case with molecular biology.

There are many differences between the restricted and unrestricted sciences. For one, their interpretation of the scientific method is not the same. But here I want to discuss their different attitudes toward an external world of 'real' objects. That such a difference exists can be seen very simply in the percentage of practical marks against theory in a recent university examination in various Natural Sciences: Geology 40, most biological subjects 33, Chemistry 30, Physics 20, Theoretical Physics 0.

The difference appears in two ways. The unrestricted sciences deal with a richer variety of phenomena than the restricted: and in particular the goals of their study may be phenomena at many different levels of size and complexity, as in the large-scale problems of the geologist, the taxonomic relationships of starfish, the machinery of the central nervous system, the conduction of the nervous impulse, the molecular-biological problems of the replication of nucleic acids, and so on. Scientific attack is based partly upon analysis of factors which bear upon a phenomenon. This can rather easily lead to what I might call the 'analytical fallacy': that understanding of a phenomenon is only to be gained by study of rules governing its component parts. Particularly in the restricted sciences we seem to see a progressive analysis starting with gross physical objects, the understanding of which depends upon molecular analysis by the chemist, which in turn depends upon our knowledge of isotopes, which in turn depends on the ever-increasing number of 'ultimate' particles dispensed to us by the nuclear physicist. From here it is easy to pass to the fallacy that once we have found the correct assumptions necessary for the description of ultimate particles we have only to work out the consequences of these, together with the theory of probability, to describe the properties of all material configurations of higher and higher orders. As Price says in his work *Perception*: 'Thus the not uncommon view that the world we perceive is an illusion and only the "scientific" world of protons and electrons is real, is based upon a gross fallacy, and would destroy the very premisses upon which science itself depends.'[5] That is a view based upon analytical fallacy. Price's statement will do well so long as we remember that it describes a common error arising from the present state of the sciences, and not the view of the informed man of science.

Now as we pass to higher orders of configurations we find new, so-called 'emergent' properties, such as the special properties of living systems which distinguish them from the non-living, or the predictor properties of brains and computing machines. Do the assumptions for ultimate particles suffice for these emergent properties? At the outset, empirically, they do not. That is, even if the physicist one day gets to some really ultimate particles, it would be long before we could extrapolate upwards in the manner required—and, in practice, novel features of complex configurations would still require new assumptions. But in fact the present position of nuclear physics suggests that the quest for ultimate particles may never reach finality. In the 1930s, Eddington could indeed suggest that the universe consisted of 10^{79} protons and 10^{79} electrons, a number bound up with the dimensions of the universe itself.[6] Later, as Heisenberg said, neutrons were added to these two components.[7] But hope was deferred. Soon after this temporary breathing space, other particles were discovered and their number already exceeds that of the 92-odd fixed elements of an earlier day. The biologist may be forgiven a doubt whether, in fact, there is an end to this particular analysis. And if that is so, where is the foundation upon which we can build a superstructure for the description of higher systems?

But that is not the main difficulty with the analytical fallacy. It is this: higher-order configurations may have properties to be studied in their own right. We can make observations to enable us to understand how a petrol engine works without calling upon the molecular hypothesis. Chemical analysis may help us to make more enduring cylinders; but that is a problem with a different goal. In the same way, electron microscope sections of the components of a computing machine will not help us to understand how it works or the origin of the highly significant parallels between the principles of its action and some of those which seem to govern central nervous action.

It is simple systems that occupy the particular attention of the restricted sciences. The unrestricted sciences deal with innumerable complex systems with seemingly emergent properties. Understanding of these is not to be obtained by extrapolation of their simpler components. What has to be done here is essentially a taxonomic operation—the determination of the class to which a phenomenon belongs. That was the key to our understanding of nervous action. Equally, I consider that the vitally important and most intractable problems of ecology and population studies can only be advanced by seeking comparisons of class with models from physical chemistry. We need a new Willard Gibbs with a biological slant.

But for our present purposes the unrestricted scientist is always deeply aware of the multitude and variety of higher-order systems, with their emergent properties. Unlike the restricted scientist, we cannot shelve the study of phenomena which seem too complex—thereby introducing a systematic bias into the treatment of phenomena in general. The multitude and reality of these higher-order systems give the biologist an immense respect for the reality of natural phenomena, as opposed to hypotheses about them.

The nuclear physicist today presents us with a world of elementary particles which seem to have nothing in common with the everydayday objects of our experience. Indeed for him these particles are not observable as things. He seems concerned only to establish relations between them which observation can show to be constantly obeyed and which thus permit successful predictions. As Michael Faraday said in 1844: 'What thought remains on which to hang the imagination of an *a* independent of the acknowledged forces?'[8]

Heisenberg refers to such particles as 'the building stones' of matter.[9] The term will do so long as we do not suppose that description of their relationships will necessarily suffice to describe the special properties of material configurations of a higher order; that they are the sole 'building stones'. But with respect to the reality of the external world the nuclear physicist leaves us only with the conclusion that the demonstrable relationships between these particles are not mere products of our own minds, but must arise from something external to us.

The basis of acceptance of the real world in the unrestricted sciences is very different. Such a scientist at work accepts absolutely the existence of a world of real objects. He does so more completely than does a physicist or a philosopher in his everyday life, or indeed than does the everyday man. The reason for this is the complete congruence between this acceptance, and this alone, with the experience and predictions in everyday life, and that in the enormous number and variety of phenomena which through his job he critically witnesses, all seem consistent with a real world of objects.

Craik, in his admirable essay on the nature of explanation, points out that one can never probe the existence of an external thing, or its obedience to a particular law, by trying to wring the truth out of a particular example. He says: 'You must vary the conditions, repeat the experiments, make a hypothesis and test it out.'[10] That is indeed a way of approaching the matter inductively. But I do not think this conscious logical procedure is the source of our conviction of the existence of external things.

Some years ago, while engaged upon the taxonomic identification of the species of certain worms, I was greatly struck by the entirely different procedure I used when, in the field, I concluded beyond doubt, 'There is a specimen of *Rhyncodemus bilineatus*', and the procedure I used in the laboratory.[11] In the latter I slowly followed a conscious logical process of identification based upon certain well-defined 'yes or no' characters of the worm's internal anatomy. In the field I instantly recognised the species of the worm. The two methods are quite different, and are subject to quite different kinds of error. Field recognition, which I have called 'aesthetic recognition', depends enormously upon past experience, much of which is not even conscious. When I see a shore-crab today, and say, 'There is a *Carcinus maenas*', the whole machinery of my perception of it is different from what it was when I was a child. In an important sense, my recognition of a specimen of *Carcinus maenas* today is only the end of a long series of all sorts of experience, unconscious as well as conscious. And the end of all this is not arrival at a logical conclusion that shore-crabs must exist, but the tacit absolute conviction that they do, through all sorts of past experience. The appearance of colours of a shore-crab or of a tomato are not basic units from which, with similar units, we can build evidence for or against the existence of a world of real objects. A 'tomato' that appeared bright red in the dark or under a sodium lamp should be a highly suspicious object to any chef. The acceptance of the external reality of such objects depends upon the whole of past experience.

That conviction of reality is enormously enhanced by the variety and indirectness of the evidence with which it is congruent. This is most strikingly shown by a study of the behaviour of insects. The simplest cellular animals, sea-anemones, jellyfish and the like, can show remarkably complex motor reactions to natural stimuli. But it is to stimuli that they react, not to the presence of objects, as in our own behaviour. Among insects, on the other hand, the matter stands very differently. The hunting wasp, *Ammophila pubescens*, digs burrows for its young.[12] It hunts over a considerable distance for spiders and caterpillars, which it paralyses and puts in its burrows. Later the eggs hatch and the young feed upon the paralysed prey. If the prey is too large to be carried by flight, the wasp will drag its prey around obstacles along the ground towards the burrow. If wasp and prey are transferred in a closed box to a new place some distance away, the prey will nevertheless be dragged towards the burrow. The wasp behaves in fact as though it had an internal model of the district round its burrow, just as Craik suggests that we, ourselves, have an internal model of the external world which we use to control and predict action. The wasp is behaving in relation, not to stimuli, but to a world of objects, and to a world of objects identical with that accepted by our own everyday naïve realism. This is carrying congruence with the world we naturally accept very far.

But the matter does not end here. Physiological study of men and animals shows that much of their behaviour can be usefully described by considering them as predatory behaviour machines. The primary need is to foretell the future. Such prophecy is possible with high probability on the assumptions:

(1) that information collected about past events can be a guide to the course of events in the future;

(2) that, however different an organism is from ourselves, and

however different the sources of information which it appears to utilise, on analysis its behaviour is completely consistent with the occurrence of the external objects and events presented to us by the naïve realism implicit in our own everyday behaviour.

Physical studies can tell us the kinds of physical and chemical information which organisms or predictor machines can receive. When we examine animals we often find that very different kinds of sensory instruments from our own are used to receive that information. Bees have colour vision, but they behave as though their colours are quite different from those we ourselves recognise.[13] Cabbage white butterflies, on the other hand, seem to have colour appreciation very close to our own.[14] Sound in insects is generally only detected at low frequencies (below the C above middle C). But crickets have an ingenious mechanical rectification device by which they can receive the high-pitched chirrup of their stridulation.[15] And yet for all these great differences in the kind of information received, the resulting behaviour remains completely consistent with a real world of the objects familiar to us ourselves. Thus the congruence between the impressions we receive and the existence of an external world of real objects is not just something inferred from our own direct observation of particular phenomena. We can seek our phenomena through far-distant and wholly unexpected channels—and the congruence of the phenomena with a real world of external objects never fails; the experience in support of this is far greater for a trained naturalist than it is even for the ordinary man.

And there is yet one more thing of interest. Charles Darwin once said:

> It is certain that there may be extraordinary mental activity with an extremely small absolute mass of nervous matter; thus the wonderfully diversified instincts, mental powers, and affections of ants are notorious, yet their cerebral ganglia are not so large as the quarter of a small pin's head. Under this point of view, the brain of an ant is one of the most marvellous atoms of matter in the world, perhaps more so than the brain of man.[16]

Even the astonishing behaviour of which von Frisch has shown the honey bee to be capable is operated through a brain of about 0·62 cu.mm., and that of a larger ant by about one tenth of this size, against the 1,600 ml. of our own.[17] Certainly, the apparent appreciation of a real world of real objects only seems to require an utterly trivial number of nerve cells compared with our own. When we consider the problems of the mind–brain relationship, perhaps a biologist concerned with the behaviour of the lower animals may be forgiven a doubt as to whether as yet we have even begun to see the questions that must be asked about our own brain and mind.[18]

I think the question we must ask is not, 'Can we find any premisses from which the existence of an external world can be proved with certainty?'; it should be, 'Why do we accept with conviction an external world of real objects?' It is important to realise that by this we do not simply refer to tomatoes and bent sticks and so on, but to something much more complicated and unique. We recognise enduring objects, like tables and mountains, liable to denudation. We also recognise objects which are open steady states, such as rivers and the ocean. These various objects exist at many levels, from that of nuclear particles up to ecological systems and to the universe itself. All behave with respect to time according to elaborate rules, from which, for example, the physical chemist can distil such statements as the second law of thermodynamics, or the biologist that of natural selection, which, like the second law, tells us something about the probable future of material configurations. Of course, the physical chemist will tell us that the second law is only truly applicable to ideal systems of a certain sort. But it must not be forgotten that it holds well enough in everyday life for it originally to have been based upon experimental observations. It must not be supposed that such well-defined rules are consciously present in our minds when we deal with everyday life: but our uninformed and indeed unconscious expectation of what will happen in the world follows these rules closely.

All this elaborate system of material objects changing with time follows a pattern consistent with past and present experience: stones follow the same rules for mice and men. One feature of these patterns is of particular importance. Phenomena fall into classes. That is both a character of the 'real world', and it is the basis of the fact that models can be made showing identical essential features, so that the behaviour of the phenomena can be predicted from that of the model. A system in relaxation oscillation can be built either electronically or hydraulically. Either can be used as a model to predict the behaviour of the other.

Behaviour patterns tacitly accepting the reality of material objects and their changes in time are to be seen in ourselves, in other men, and, as I have said, in quite lowly animals. Our recognition of shore-crabs and tomatoes is based upon the consistency of that recognition with the whole of past experience. On occasion we may make errors or suffer hallucinations: but it is only a question of time before these come into collision with the expectations of experience. Often such errors are due to incomplete present information leading to wrong classification. When the conjuror saws the lady in two, all experience leads us to suppose that behind the scenes we should 'see how it was done'.

I think that it is this whole consistent pattern of things and their changes in time which engenders the tacit acceptance of reality of the external world. Particularly for the biologist, who observes that even lowly creatures behave as though they tacitly accept the same external world as we do, the question arises as to whether our own acceptance is a conscious process at all. If there is a square-topped table in the room I react appropriately to it whether or not I am consciously aware of it. If questioned I may consciously note features of it, colour, shape and so on. But it does not follow that such features as I can consciously perceive about it are the essential basis of my conviction of its reality. That conviction arises from the whole of my past experience, conscious and unconscious, and the consistency of the phenomena which it presents; and my everyday acceptance of the reality of the external world depends particularly upon the unconscious assumption that the present kind of consistency will not suddenly fail. The past would then be no guide to the future, and the basis of any such unconscious assumption would collapse. It is impossible to prove that that failure might not occur, for all our prediction of the future depends upon past experience. Only on the assumptions implicit in that can we form an inductive proof of the reality of the external world. There can be no deductive proof.

As a biologist, it seems to me that the problem of our acceptance of external reality has often been complicated by concentration upon conscious perception. Thus, Dr Price in his book on 'Perception' begins:

> Every man entertains a great number of beliefs concerning material things, e.g., that there is a square-topped table in this room, that the earth is a spheroid, that water is composed of hydrogen and oxygen. It is plain that all these beliefs are based on sight and touch (from which organic sensation cannot be separated): based upon them in the sense that if we had not had certain particular experiences of seeing and touching, it would be neither *possible* nor *reasonable* to entertain these beliefs.[19]

He then goes on

> . . . to examine those experiences in the way of seeing and touching upon which our beliefs concerning material things are based, and to inquire in what way and to what extent they justify these beliefs. Other modes of sense experience, e.g. hearing and smelling, will be dealt with only incidentally. For it is plain that they are only auxiliary. If we possessed them, but did not possess either sight or touch, we should have no beliefs about the material world at all, and should lack even the very conception of it.[20]

In the first place, as I have said, it does not seem to me proven that my assent to the existence of such things as a square-topped table is purely the result of conscious perception. It seems possible to consider that what are commonly referred to as sense-data are not the elements from which our assent to the existence of an object is derived, but rather that they are to be considered as labels of which we can become consciously aware, and which are attached to certain kinds of information we receive about an object.

Secondly, the statement about sight and touch does not seem to me to be true. A man blind from birth can have all the beliefs to which Dr Price refers. Touch is an exceedingly complex and ill-defined sense. It is worth bearing in mind the physiologist's view. Winton and Bayliss, reviewing the effects of cortical lesions, say:

In man, the destruction of the sensory area does not abolish sensations of pin-prick, touch, heat or cold. It does diminish the power of localising a stimulus sharply and appreciating accurately fine differences. Stereognosis, the power of recognising the shape of an object when it is held in the hand, is always severely impaired in these lesions. The recognition of an object by touch, which seems childishly simple to a normal subject, requires sensations of touch, pressure, joint and muscle sense, the fusion of the separate sensory data, and the recollection of previous similar experiences.[21]

Though it is dangerous to isolate and commend the relative importance of any of the senses, the importance of hearing and smelling should not be belittled—particularly when extended to the lower animals.

The really important question is: are there unconscious sources of information which contribute to our behavioural assent to a real world of external objects? A comparative physiologist who studies the behaviour—and the powers of intercommunication—of bees and ants is at least forced to be aware of this question. And in ourselves, when we drive an automobile correctly through a maze of traffic lights, it is hard to suppose that we consciously perceived each—even though some of them could subsequently be recalled to consciousness. Vision itself may be unconscious as well as conscious, and since in both cases behaviour is affected, the word perception itself needs qualification.

But most interesting of all are those classes of sensory information which undoubtedly contribute to knowledge of the world around us unaccompanied by sense-data. Such senses are particularly the sense of orientation associated with the inner ear and above all that of proprioception. That 60-year-old term of Sherrington's, proprioception, has crept into the supplement of the 12-volume *Oxford English Dictionary*, though older and imprecise terms such as 'kinaesthetic sense' are well established even in the smaller brethren of the great work.[22] Since proprioception is one of the fundamental necessities of animal life its importance should be appreciated. By orientation and proprioception an organism is aware of its position in space and the relationship of its parts. We cannot assign sense-data to proprioception in the way we can conceive of patches of redness in vision. Position and orientation seem simply inherent in our parts. Yet, notwithstanding Dr Price's statement, a good case could be made for supposing proprioception even more important than vision to our primary assented notions about space and its physical objects. All the visual difficulties of telling whether a stick partly immersed in water is straight or bent are overcome by running a hand down it, even in the dark and when our fingers are numb with cold.

Conscious perception therefore at best provides only part of the information contributing to our notions of the external world. It is noteworthy that we owe it to the physiologist that this has been brought to our notice. At times, the attempt is made to exclude the physiologist from discussing matters of this sort on the grounds that his experiments presuppose the very things at issue. But this is scarcely right, since he has in fact repeatedly drawn attention to possibilities which have been overlooked: and too often the attempt to evade him only succeeds in an unconscious return to the physiological premisses of an earlier day. The five senses that hold sway in so much discussion are merely the supposed physiological sources of information of two hundred or more years ago.

This still leaves us with the interesting question of why it is that only certain sources of information about the external world are accompanied by sense-data labels. I have no answer to this except to note that our orientation and position are not good taxonomic features of the objects in the world; whereas consciously seen red patches, musical notes, odours, taste and touch, are exactly the kind of taxonomic features which, fed into a digital computing machine, could deliver to us far-reaching reasoned logical conclusions.

Notes and references

Chapter 1: The restricted and the unrestricted sciences: pp. 1–25

1 J. Ray, 1735, Preface to *The Wisdom of God as manifested in the Works of the Creation*, London, Innys and Manley.
2 *Chambers's Encyclopaedia*, 1950, **12**, 280, London, Newnes.
3 J. Playfair, 1812, *Outlines of Natural Philosophy*, 1, 1 and 2, Edinburgh, A. Constable.
4 S. H. Mellone, 1902, *Introductory Textbook of Logic*, 12th edition, 1920, p. 293, London and Edinburgh, Blackwood.
5 W. S. Jevons, 1870, *Elementary Lessons in Logic*, 1905 reprinted, p. 249, London, Macmillan & Co.
6 K. Lucas, 1910. On the recovery of muscle and nerve after the passage of a propagated disturbance. *J. Physiol.* **41**, 268–408.
7 E. D. Adrian, 1912. On the conduction of subnormal disturbances in normal nerve. *J. Physiol.* **45**, 389–412.
8 A. Eddington, 1933, *The Expanding Universe*, 1952 reprinted, pp. 68, 113. Cambridge University Press.
9 C. B. Travis, 1914. Presidential Address. *Proc. Lpool geol. Soc.* **12** (pt. 1), 1–31.
10 A. C. Ramsay, 1846. On the denudation of S. Wales and the adjacent counties of England. *Mem. geol. Surv. U.K.* **1**, 297–335.
11 O. T. Jones, 1924. The Upper Towy Drainage System. *Q. Jl geol. Soc. Lond.* **80**, 568–609.
12 L. Agassiz, 1842. The Glacial Theory and its recent progress. *Edinb. New Philos. J.* **33**, 217–40.
13 W. S. Jevons, 1870, *op. cit.* pp. 231, 232.
14 C. Darwin, 1903, *More Letters of Charles Darwin*, ed. F. Darwin and A. C. Seward. 1, 195, London, John Murray.
15 Pantin is here approaching from a rather different angle a problem which has recently been discussed by Michael Polanyi (*The Tacit Dimension*, London, Routledge and Kegan Paul, 1967). Polanyi points out that a machine can be described as a particular configuration of solids. The description would state the materials and shapes of the parts, and the boundary conditions by which they are joined together as a system. But this could describe only one particular specimen of one kind of machine. It could not characterise a class of machines of the same kind, which would include specimens of different sizes, often with different materials, and

with an infinite range of other variations. Such a class would be truly characterised by the operational principles of the machine, including the principles of its structure. It is by these principles, when laid down in the claims of a patent, that all possible realisations of the same machine are legally covered; a class of machines is defined by its operational principles. Polyani goes on to observe: (1) that a particular specimen of a machine is characterised by the nature of its materials, by the shape of its parts and their mutual arrangement which can be defined by the boundary conditions of the system; and (2) that the laws of physics and chemistry are equally valid for all solids, whatever their materials and shapes and whatever the boundary conditions determining their arrangement. He argues that machine-like functions, whether in machines designed by human beings or in the machine-like structure seen in living organisms, are based on what can be called the principle of marginal control. Borderline conditions are left open by operations of a lower level (in this case physics and chemistry), enabling the higher level (in the case of human machines, the aims of the designer) to control these borderline conditions which are left open by the operations of inanimate matter. Engineering provides a determination of such borderline conditions, and this is how an inanimate system can be the subject of dual control at two levels: the operations of the upper level are artificially embodied in the lower level, which is relied on to obey the laws of inanimate matter.

16 A. N. Whitehead, 1920, *The Concept of Nature*, 1955 reprinted, p. 82, Cambridge University Press.

17 In writing the summary for this chapter, which summary is of course placed in its proper place at the end of the chapter, Pantin had added half a page of additional material summarising matter which in fact he left out of the lectures as delivered. It seems useful to include it here as a footnote, and this is what I have accordingly done (W.H.T.).

'When considering the relation of our goals of scientific investigation to the phenomena observed, it is useful to consider the relation of the phenomena to Whitehead's "sets of extensive abstraction". He shows that such abstractive sets are the basis for the ideal timeless instants and dimensionless points of the mathematical models by which we represent some kinds of reality. But it is important to remember that each element of a set is not a "homogeneous substance". It is an actual event in space and time. Such events may have a significance in themselves at any stage of abstraction which is lost by further abstraction—as with the performance of a play in a particular theatre. The event which is the whole performance of that play loses significance (that is, a certain extension of knowledge and understanding about our world) if we abstract a brief arbitrary portion of the performance. Therefore, in abstracting events, we can find particular abstractive sets which show significance at various levels of abstraction. It is a feature of perceptive experience that this is possible: the world can be considered in such a way that it "makes sense". It is at such various and particular levels that we find the goals of scientific investigation.'

Chapter 2: The features of the natural world: pp. 26–52

1 It is possible by appropriate brain stimulation to cause a domestic fowl, for example, to behave as if suffering from a hallucination. Many birds, of which the domestic fowl is one, will show strong aggressive response to an object such as a model, or stuffed example, of a weasel or an owl. If the brain of the bird is electrically stimulated in the appropriate place a response to such dummies will be greatly increased and the bird may launch out into a full-scale attack—a course of action which it will not otherwise carry to such an extreme. Still more remarkable is the fact that a higher intensity of stimulation in the same place will, even in the absence of any stuffed or dummy weasel at all, cause the fowl to behave as if it were attacking a weasel though there is none there. In this case the fowl has every appearance of being the victim of a hallucination, for it first attacks a spot in empty space and finally flaps away from 'nothing', screaming with fright! Examples such as this provide evidence for the basic similarity between the perceptual processes of man and of many higher animals. But animals can also come to 'realise the unreality of hallucinations'. Thus a dog may at first be deceived by a picture, a mirror image or a sound recording; but then suddenly, without any sign of 'trial-and-error' behaviour, 'realise' its mistake and thereafter pay no attention to such things. (E. von Holst and U. von St Paul, 1960, 'Vom Wirkungsgefüge der Triebe', *Naturwissenschaften*, **47**, 409–22; for a fuller account see W. H. Thorpe, *Science, Man and Morals*, 1965, London, Methuen, pp. 108–9.)

2 Here, and in what follows, Pantin is approaching the coherence theory of truth from the biological viewpoint. For a valuable critique of the 'coherence theory' as against the empiricist or 'correspondence theory' of truth, a critique which supports Pantin's discussion, see E. E. Harris, 1954, *Nature, Mind and Modern Science*, London, Allen and Unwin, especially pp. 197–200 and 352–5.

3 M. von Senden, 1960, *Space and Sight: The perception of space and shape in*

the congenitally blind before and after operation, London, Methuen (translation of *Raum und Gestaltfassung bei operierten Blindgeborenen*, 1932).

Von Senden hoped that study of his cases would yield information about the nature of perceived space, but his subjects were too varied in age, sex, intellect and background for generalisations from them to be possible, nor did his study reveal much about the importance of early visual training for perception in the adult. It did show, however, that many patients find it extremely difficult to interpret their new visual experience, taking months to attain any proficiency, and that some never succeed in doing it at all.

A recent monograph by Gregory and Wallace sets forth a case history of much interest which, although it does not have much relevance to the theory of perception, does demonstrate that early touch-experience may link up with vision many years later. This patient was a man who had been effectively blind almost since birth and recovered appreciable sight by operation at the age of 52. He rapidly learned to link up visual experience with previous touch-experience, for instance to read capital letters which he had learned by touch from wood blocks, but was much slower—as one would expect—to interpret visual experience unconnected with touch, e.g. he was slow to learn to read lower-case letters which he had never handled, and was baffled by the expressions on faces.

Perhaps the most interesting aspect of this case is the psychological effect of the patient's altered image of himself. While he was blind, his self-image was of a 'superior' person, a conqueror overcoming severe disadvantages with energy and resolution. When his sight was restored and he had, in theory, the same equipment for life as everybody else, he became aware of his relative inadequacy, that he was an 'inferior' person, his self-image was impaired, and he became deeply disturbed and depressed.

A full account of this interesting case is given in the following paper: R. L. Gregory, and J. G. Wallace, 1963. Recovery from early blindness: a case study. *Exp. Psych. Soc. Monograph No. 2.* Cambridge.

4 H. H. Price, 1932, *Perception*, London, Methuen. See p. 3 of 1964 edition.

5 Those interested in the organisation of perception in animals and in particular the question of perceptual relations and the formation of gestalten will find a discussion of this subject in W. H. Thorpe, *Learning and Instinct in Animals*, 2nd edition 1963, chapter 7, pp. 138–47.

6 E. H. Land, 1959. Color vision and the natural image. *Proc. U.S. Nat. Acad. Sci.* **45** (1), 115–29, and (4), 636–44.

7 In the manuscript Pantin had here a pencilled note referring to the small

volume of nervous matter in certain animals. His own later thought on this will be found in his paper entitled *Learning, World-Models and Pre-adaptation* which is printed as appendix 2 at the end of this book.

8 For further account of the colour vision of bees and the importance of ultra-violet in bee perceptions, see K. von Frisch, 1927, *Aus dem Leben der Bienen*, Berlin. Also *The Dance Language and Orientation of Bees*, 1967, Cambridge, Mass.: Harvard Press.

9 E. Schrödinger, 1944, *What is Life?* Cambridge University Press. See pp. 70 and 71.

10 The importance of Le Chatelier's rule as applied to animal behaviour is shown by George Humphrey's utilization of it in his pioneer work on the type of learning known as 'habituation'. Used in its widest sense, habituation is a simple learning not to respond to stimuli which tend to be without significance in the life of the animal. It is thus a tendency to drop out responses, not to incorporate new ones or complicate those already present. In this respect it is certainly the simplest kind of learning, and something like it is universal in animals. Perhaps, indeed, in one or other of its forms it may be said to be one of the fundamental properties of living matter. (G. Humphrey, 1930. Le Chatelier's rule and the problem of habituation and de-habituation in *Helix albolabris*. *Psychol. Forsch.* **13**, 113–27; G. Humphrey, 1933, *The Nature of Learning in its Relation to the Living System*, London, Kegan Paul.)

11 J. H. Woodger, 1960. Biology and Physics. *Br. J. Phil. Sci.* **11**, no. 42, p. 89.

12 For a discussion of the concept of mechanism and the extent to which the word 'machine' can be used to describe living organisms and their parts see M. Polanyi, 1967, *The Tacit Dimension*, London, Routledge and Kegan Paul.

13 For Pantin's further thoughts on some of the topics considered in this chapter see appendix 1, *Life and the conditions of existence*, being a paper contributed to the volume *Biology and Personality*, edited by I. T. Ramsey, 1965, Oxford, Blackwell.

14 A brief but profound and beautifully written essay dealing with the relation between the world as perceived by the scientist and the reality which lies behind it, an essay which Pantin might well have referred to had he lived to produce this book himself, is that by Sir Cyril Hinshelwood, *The Vision of Nature*, 1961, Cambridge University Press (being the 15th Eddington Memorial Lecture).

Chapter 3: Living systems and natural selection: pp. 53–76

1 W. Whewell, 1840, *Philosophy of the Inductive Sciences*, 2nd edition, 1847, London, John Parker; see I, 602.

2 T. H. Huxley, 1868, *On the physical basis of life*. Collected essays of T. H. Huxley (1893), I. Method and results. London, Macmillan.

3 Lord Kelvin, 1884. Hathaway's Papyrographed Edition of Lectures given at Johns Hopkins, 1883, p. 132.

4 Clerk Maxwell, 1875, *Encyclopaedia Britannica*, article on Atom.

5 R. Chambers, 1844, *Vestiges of the Natural History of Creation*, London, John Churchill.

6 K. Pearson, 1892, *The Grammar of Science*, Everyman Ed. 1937; see p. 282.

7 A. Weismann, 1889, *Essays upon Heredity and Kindred Biological Problems*, 2nd edition 1891, Oxford, Clarendon Press.

8 Directiveness *versus* Purposiveness. From the philosophical point of view, the central problem of ethology is the relation between purposiveness ('purpose' here has the usual meaning—a striving after a future goal retained as some kind of an image or idea) and directiveness, and it looks at times as if this is the same as the relation between learning and instinct. All biologists agree that the behaviour of organisms as a whole is directive, in the sense that in the course of evolution some at least of it has been modified by selection so as to lead with greater or less certainty towards states which favour the survival and reproduction of the individual. All machines are also directive in the sense that their parts have been designed or selected so as to behave in a particular way whenever activated by an external source of power; but not even the most elaborate machine, such as an electronic calculator, is purposive. So for the ethologist the question is, 'How much, if any, of the animal's behaviour is purposive and what is the relation of this behaviour to the rest?'

In human perception, as H. M. Price (1932, *Perception*, London) has shown, the very idea of a material object is dependent upon an element of anticipation. He says 'every perceptual act anticipates its own confirmation by subsequent acts'. A. N. Whitehead (1929, *Process and Reality*, Camridge) considers the act of perception as the establishment by the subject of its causal relation with its own external world at a particular moment. Whitehead argues that every vital event, in fact, involves a process of the type which, when we are distinguishing between mental and material, we describe as mental—the act of perception. A very strong case is made by W. E. Agar (1943, *The Theory of the Living Organism*, Melbourne University Press) for the theory that a living organism is essentially something which perceives. Therefore some element of anticipation and memory, in other words *some essential ability to deal with events in time as in space is, by definition, to be expected throughout the world of living things.*

9 With regard to the improbability of complex machines M. Dixon and E. C. Webb (1964, *Enzymes*, 2nd edition, London, Longmans) have some very relevant remarks on the crucial question of information transfer and storage in the intracellular mechanism for protein synthesis. Thus, they say:

> The code is built into the biosynthetic mechanism in two places: (*a*) it is embodied in the DNA of the genes, which use it to represent the corresponding proteins, and (*b*) it is built into the specificity of the ligases. One of the most interesting and fundamental questions in biology is how it comes about that the genes and the ligases use the same code. The answer that no life is possible unless they do is unsatisfying; the chances against it coming about without some controlling mechanism to relate the two are enormous, but it is extremely difficult to picture such a mechanism.

And again:

> The structure of both the specific centres in an enzyme which is subject to feedback inhibition are determined by its structural gene. The genes indeed display an astonishing amount of 'knowledge' about the sequence of chemical processes in metabolism. One may well ask how the gene-forming enzyme 2.4.4.14 'knows' that phosphoribosyl pyrophosphate will be converted by the consecutive action of ten or more different enzymes into a purine nucleotide, or how the gene for the first enzyme of histidine biosynthesis, which acts on the same compound, 'knows' that its product will be converted into histidine by a different series of enzymes. Even with this information, how do these genes 'know' what amino acid sequences in their enzymes will act as specific centres combining with purine nucleotides or histidine respectively? Evidently there must be some mechanism whereby information derived from the metabolic processes themselves is transmitted back to the genes and there incorporated in the form of polynucleotide sequences. The manner in which control was established in the first place, and the nature and mode of action of this mechanism, is one of the most fascinating and fundamental questions in biology.

10 DNA, in so far as it is the carrier of information, is not regarded as a machine by Polanyi (1967) since the code depends upon a 'read-out' by a particular kind of mechanism; the information depends on this taking place starting at the right point and reading in the right direction. Thus the structure of the DNA from this point of view is only machine-like

in the sense that the arrangement of type on the page of a book is a machine for conveying information. The marks on the paper are only meaningful if there is a 'read-out' mechanism capable of recognising and grouping them in the right manner.

11 L. J. Henderson, 1913, *The Fitness of the Environment*, New York, Macmillan.

12 P. M. A. Broda, 1960. The mechanism of inheritance. *The Adv. of Sci.*, **16**, no. 64, p. 339.

13 The concept of a self-duplicating molecule which contains all the information, and to which protein can contribute none, has been criticised by Commoner (1962, Is DNA a self-duplicating molecule?, in *Horizons in Biochemistry*, ed. M. Kasha and B. Pulman) and by A. C. R. Dean and Sir Cyril Hinshelwood (1963, *Nature, Lond.* **199**, 7–11; **201**, 232–9; **202**, 1046–), who argue that DNA is not merely self-duplicating; there is still a possibility that enzymes and proteins may be contributing information and may be essential to the DNA reproductive process. If this is so, the question arises whether one can have a minute-scale store without an enormous redundancy, since even if the DNA is resistant to thermal noise, the other components of the duplicating mechanism will not be so.

14 Pantin is here discussing the question, which has been referred to above (see note 10), whether or not the information for an organism is all, or primarily, in the DNA molecule, or whether that molecule can only be regarded as providing information in the environment of the cell—which view seems to carry the corollary that a substantial amount of information is being supplied by the cell itself. In what follows the author seems to come to the conclusion that it is impossible to separate the information of the DNA from that in the cell, for he says that no particle of matter nor any event is wholly uninfluenced by the rest of the natural world. See also the suggestion above (note 8) that 'information' itself cannot be legitimately regarded as being mechanistic.

15 M. Faraday, 1844. On Electric Conduction and the Nature of Matter. *Philos. Mag. and J. of Sci.* **24**, 136–44.

16 J. Tyndall, 1868, *Faraday as a Discoverer*, London, Longmans Green, pp. 123–4.

17 A. N. Whitehead, 1925, *Science in the Modern World*, Mentor ed. 1958, pp. 149–50.

18 W. H. Thorpe has attempted to discuss this problem of the concept of emergence in the light of recent scientific and philosophical writing on the subject in *Science, Man and Morals*, 1965, London, Methuen, pp. 20–6.

19 G. H. Lewes, 1875, *Problems of Life and Mind*, I, 98, London, Trubner.

20 C. Lloyd Morgan, 1933, *The Emergence of Novelty*, London, Williams and Norgate; see p. 59.

21 In this passage Pantin comes to the core of the argument and arrives at much the same conclusion as Polanyi, who continually warns against the dangers of what he calls 'pseudo substitution'. This is the common mistake (under the plea of ignorance as to what material components of a system really are) of putting what is to be explained surreptitiously into the concepts which are to serve for explanation. This is a pit into which C. H. Waddington (1961, *The Nature of Life*, London, Allen and Unwin) seems to fall in his discussion of what he called 'the new vitalism'. It was perhaps the greatest of the great contributions of Whitehead to show in his *Philosophy of Organism* that even the antithesis atomism *versus* continuism is a false one, and that perhaps after all it is organicism that is fundamental. If this is so, then fundamental biology is the study of genes, cells, tissues and whole organisms and their behaviour. The most highly developed relationships that living things exhibit are just as fundamental, perhaps more fundamental for science, as the idea of the ultimate atomic unit. He summed all this up in his famous aphorism, 'Biology is the study of the larger organisms and physics the study of the smaller ones.' But here too we must be on our guard against lightly supposing that the 'organicism' of the smaller units is in any way such as to 'account for' that of the larger.

22 See E. E. Harris, 1965, *The Foundations of Metaphysics in Science*, London. On page 232 Harris discusses a point raised by a number of writers and in particular summarised by L. L. Whyte (1965, *Internal Factors in Evolution*, London, Tavistock Publications) that the conditions of biological organisation restrict the possible avenues of evolutionary change from a given starting point. The nature of life in fact limits its variation and is one factor directing phylogeny. Harris, quoting both T. H. Morgan and L. T. Hogben to the effect that selection (both internal and Darwinian) produces trends by canalising and restricting life to certain combinations from amongst the infinity theoretically possible, says that Morgan and Hogben were right in saying that if no selection had occurred all the known forms of life would still have appeared as well as an enormous number of others. But of course, to produce these an infinitely long time would have been required. See also R. A. Fisher, 1950, *Creative Aspects of Natural Law*, being the 4th Arthur Stanley Eddington Memorial Lecture, Cambridge University Press.

Chapter 4: The classification of objects and phenomena (pp. 77–99)

1 W. Whewell, *op. cit.* I, 468.
2 W. C. T. Calman, 1919. On barnacles of the genus *Megalasma* from deep-sea telegraph cables. *Ann. Mag. Nat. Hist.* (9), **4**, 361–74. See also W. C. T. Calman, 1949, *The Classification of Animals: an introduction to zoological taxonomy*, London, Methuen, pp. 27–8, 37–8.
3 C. E. Raven, 1942, *John Ray: Naturalist. His Life and Works*, Cambridge University Press.
4 John Ray, 1674, *The specific differences of plants*, paper presented to the Royal Society, November 1674. See T. Birch, 1775, *History of the Royal Society of London*, III, 169, London, A. Millar.
5 W. Whewell, *op. cit.* I, 493.
6 C. T. Regan, 1926. *Brit. Assn. Adv. Sci. Reports*, 1926, p. 75 (Presidential address, Section D, British Association, Southampton 1925).
The exact quotation is: 'A species is a community...whose distinctive morphological characters are, in the opinion of a competent systematist, sufficiently definite to entitle it...to a specific name.' In these pages Pantin is giving an account of earlier views on what is often named Gestalt Perception. A valuable but little-known book which contains chapters dealing with this from many different aspects—physical, astronomical, biological, botanical, biochemical, embryological, neurological, psychological and artistic—is *Aspects of Form: a symposium on form in nature and art*, edited by L. L. Whyte, 1951, London, Humphries.
7 W. Whewell, *op. cit.* I, 475.
8 R. Owen, 1849, *Address to the Royal Institution on The Nature of Limbs*, London, John van Voorst.
9 A. J. Salle, 1939, *Fundamental Principles of Bacteriology*, 4th edition, 1954, p. 401, London, Baillière, Tindall and Cox.
10 C. F. A. Pantin, 1951. Organic Design. *Advmt Sci., Br. Ass.* **8**, 138–50. See also appendix 1 in this book.
11 K. Lucas, *J. Physiol.* **41**, 268–408.
12 E. D. Adrian, *J. Physiol.* **45**, 389–412.
13 W. M. Bayliss, 1915, *Principles of General Physiology*, London, Longmans.

Chapter 5: Methods in the unrestricted sciences: pp. 100–122

1 R. A. Hinde has pointed out (1956, Ethological models and the concept of drive. *Br J. Phil. Sci.* **6**, 321–31) that models in biology are most likely to be fruitful in promoting further advance and illuminating experiment in so far as they are sufficiently concrete to be related to the types of mechanism

which are, at the given stage of the science concerned, capable of being studied in organisms, while at the same time sufficiently abstract to reveal new possibilities and new quantitative relationships. See also R. A. Hinde, 1960, Energy models of motivation, *Symp. Soc. exp. Biol.* no. 14, *Models and Analogues in Biology*, ed. J. W. L. Beament, Cambridge University Press, pp. 199–213.

2 J. Loeb, 1918, *Forced Movements, Tropisms and Animal Conduct*, Philadelphia, Lippincott.

3 P. Ullyot, 1936. The behaviour of *Dendrocoelum lacteum*. II. Responses in non-directional gradients. *J. exp. Biol.* **13**, 265–78.

4 J. Muller, 1843, *Elements of Physiology*, London, Taylor and Watson, p. 640.

5 W. Whewell, *op. cit.* 1, 601.

6 W. Nernst, 1899. *Gött. Nachr. Math. Phys. Kl.* p. 104. 1908. Zur Theorie des elektrischen Reizes. *Arch. f. d. Ges. Physiol.* **122**, 275.

7 A. V. Hill, 1910. A new mathematical treatment of changes of ionic concentration in muscle and nerve under the action of electric currents, with a theory as to their mode of excitation. *J. Physiol.* **40**, 190–224.

8 K. Lucas, *J. Physiol.* **41**, 268–408.
E. D. Adrian, *J. Physiol.* **45**, 389–412.

9 L. Lapicque, 1926, *L'Excitabilité en fonction du temps*, Presses Universitaires de France.

10 In a stimulating article T. H. Bullock (1966, Strategies for Blind Physiologists with Elephantine Problems; being the Prologue, pp. 1–10, of *Nervous and Hormonal Mechanisms of Integration. Symp. Soc. exp. Biol.* no. 20, ed. G. M. Hughes, Cambridge University Press) has, in discussing the tremendous developments of neurophysiology in recent years, pointed out that 'a whole world lies before us, of integrative units that receive converging inputs, process information according to weighting factors, transfer functions, and network connexions, to achieve an abstraction of certain qualities from the arriving messages, in other words recognition on predetermined criteria. The types of units in the optic nerve of the frog are not the same as those so far described in goldfish; they are similar to but not identical with those in the pigeon; they are quite different from those in the cat, which in turn differ from units in the rabbit. That is one stage in one pathway, the optic nerve; we have all the stages and all the pathways!' See also T. H. Bullock, 1959. The Neuron Doctrine and Electrophysiology. *Science*, **129**, 996–1002.

11 W. Grey Walter, 1959. Mirror for the Mind. *Listener*, **62**, no. 1592, p. 521.

12 An elegant example of the process of abstraction, filtering and recognition by nerve cells is given by N. Suga, 1965. Functional properties of auditory

neurons in the cortex of echo-locating bats. *J. Physiol.* **179**, 26–53, who records the discovery of units in the bat's auditory cortex which will not fire to any pure tone but only to a tone which is frequency-modulated in the descending direction and in a certain frequency range. The beauty of this case is that such units can be explained plausibly by convergence of two simpler units of types commonly found at lower levels.

13 Hence the motto of the Royal Society, *Nullius in Verba.*

14 I. Newton, 1728, *The Chronology of Ancient Kingdoms Amended*, introduction, p. 8, London, J. Jonson.

15 J. Ussher, 1660, *Chronologia Sacra*, ed. T. Barlow, Oxford.

16 C. Lyell, 1830–3, *Principles of Geology*, London, John Murray.

17 W. Thomson (Lord Kelvin). 1866. 'The Doctrine of Uniformity' in Geology briefly refuted. *Proc. R. Soc. Edinb.* v, 512–13.

18 T. H. Huxley, 1869. Presidential Address, Geol. Soc. *Proc. Geol. Soc.*, **19**, 2.

19 S. T. Coleridge, 1817, *Biographia Literaria*, chapter 14, para. 2, London, Fenner.

20 J. H. Newman, 1870, *An essay in aid of the grammar of assent*, new edition (Harrold) 1947, London, Longmans Green.

21 C. F. A. Pantin, 1944. Terrestrial nemertines and planarians in Britain. *Nature, Lond.* **154**, 80.

C. F. A. Pantin, 1951. Organic design. *Advmt Sci. Br. Ass.* **8**, 138–50.

C. F. A. Pantin, 1961. Geonemertes: a study in island life. Presidential address to the Linnean Society, 24 May 1960. *Proc. Linn. Soc. Lond.* **172**, pt. 2, 137–52.

22 A. N. Whitehead, 1948, *Science and the Modern World*, New York, Macmillan (Mentor Edition 1958, p. 21).

23 O. Köhler, 1950. The ability of birds to 'count'. *Bull. Anim. Behav.* **9**, 41–5.

24 W. H. Thorpe, 1963, *Learning and Instinct in Animals*, London, Methuen (2nd edition, pp. 391–2).

25 In this connexion it is perhaps worth recalling the remarkable behaviour of the scout bee when returning to the colony which has swarmed and indicating to it, by the usual method of dancing, the merits of a home for the swarm which it happens to have discovered in its reconnaissance. Here the scout bee has to indicate, not the direction and distance of a particular food supply, but the direction and distance of a hollow tree or other similar shelter, and it is important that the swarm should ultimately go to that new home which, amongst all those discovered by the scout bees, is most suitable for its needs. Therefore the dance of the scout bee has in this instance to indicate upon a uniform scale the merits of the home discovered. M. Lindauer (1961, *Communication amongst Social Bees*,

Cambridge, Mass., Harvard University Press) has shown that the bee is able in some way not at all understood to sum up all the qualities of the shelter it has found, e.g. size, aspect, dryness, insulation, etc., and express these in a unitary manner by the persistence and vivacity of its dance on return. All the scout bees are thus expressing on the same scale when they return, and by the same symbolic method of communication, the merits of the shelter they have discovered. The result is that those bees which show by their dances that they have discovered a site of good quality attract more followers, who then, themselves, go out and inspect the site and in their turn return and dance. Those who have discovered a less attractive site dance with less vivacity and persistence, and thus secure fewer followers to go and investigate the site. Thus, before long, a majority of the bees will be advertising the direction and distance of the best site, so that in due course the swarm will in fact leave for that site. Lindauer was, however, able to show experimentally that if two sites of exactly equal merit were discovered in opposite directions there was a considerable chance that the swarm would split into two—one half going to one site and the other to the other. The result of course was that only one of the swarms would contain the queen. The way in which the standard of communication is constantly maintained so that the messages of all the different dancers are expressed according to a common standard is still by no means understood.

26 For a summary of Pantin's work on the coelenterates see (*a*) 1952, Behaviour Patterns in Lower Invertebrates, *Symp. Soc. exp. Biol.* no. 4, *Physiological Mechanisms in Animal Behaviour*, 1950, ed. J. S. Danielli and R. Brown, pp. 174–195; and (*b*) 1952, The Elementary Nervous System, being the Croonian Lecture for 1952, *Proc. R. Soc.* B, **140**, 147–68.

27 K. Lucas, *J. Physiol.* **41**, 268–408.
E. D. Adrian, *J. Physiol.* **45**, 389–412.

28 H. Poincaré, 1908, *Science et méthode.*

29 Agnes Arber, 1954, *The Mind and the Eye: a study of the biologist's standpoint*, Cambridge University Press.

Appendix 1: pp. 129–154

1 R. Owen, 1848, *The Archetypes and Homologies of the Vertebrate Skeleton*, London, John van Voorst.

2 C. Darwin, 1859, *The Origin of Species*, reprint of the sixth edition, The World's Classics, 1956, pp. 217 and 81, London, Oxford University Press.

3 R. A. Fisher, 1930, *The Genetical Theory of Natural Selection*, p. 15, Oxford, Clarendon Press.

4 C. F. A. Pantin, 1964, *Homeostasis and the Environment. Symp. Soc. exp. Biol.* no. 18, *Homeostasis and the Feedback Mechanisms*, pp. 1–6. Cambridge University Press.

5 Cuvier, 1817, *Le Règne animal*, I, 6, Paris, Deterville.

6 M. Faraday, 1844. A Speculation Touching Electric Conduction and the Nature of Matter. *Phil. Mag.* 3rd ser. **24**, 136–44.

7 G. H. Lewes, 1875, *Problems of Life and Mind*, I, 98, London, Trubner.

8 C. F. A. Pantin, 1965, Learning, world-models and pre-adaptation, *Animal Behaviour Supplement* no. 1, pp. 1–8, London, Baillière, Tindall and Cox.

9 C. F. A. Pantin, 1951. Organic Design. *Advmt Sci., Br. Ass.* **8**, 138–50.

Appendix 2: pp. 155–170

1 C. Darwin, 1895, *Insectivorous Plants*, New York, Appleton and Co.

2 L. Cuénot, 1932, *La Genèse des espèces animales*, Paris, Felix Alcan.

3 C. F. A. Pantin, 1951. Organic design. *Advmt Sci., Br. Ass.* **8**, 138–50.

4 W. H. Thorpe, 1956, *Learning and Instinct in Animals*, London, Methuen.

5 H. S. Jennings, 1915, *Behavior of the Lower Organisms*, New York, Columbia University Press.

6 F. J. J. Buytendijk, 1919. Acquisition d'habitude par des êtres unicellulaires. *Archs néerl. Physiol.* **3**, 455–68.

7 K. Lucas, 1917, *The Conduction of the Nervous Impulse*, London, Longmans Green and Co.

8 C. F. A. Pantin, 1950, Behaviour patterns in the lower invertebrates. *Symp. Soc. exp. Biol.* no. 4, pp. 175–93. Physiological mechanism in animal behaviour. Cambridge University Press.

9 C. F. A. Pantin, 1956. The origin of the nervous system. *Pubbl. Staz. Napoli*, **28**, 171–81.

10 C. F. A. Pantin, 1954. Biology and the scientific method. *Anais Acad. bras. Cienc.* **26**, 215–18.

11 C. F. A. Pantin, 1952. The elementary nervous system. *Proc. roy. Soc.* B, 140, 168.

12 L. M. Passano and C. B. McCullough, 1963. Pacemaker hierarchies controlling the behaviour of Hydras. *Nature, Lond.* **199**, 1174–5.

13 W. H. Thorpe, 1950. A note on detour experiments with *Ammophila pubescens* Curt. *Behaviour*, **12**, 257–63.

14 R. Lorente de Nó, 1938. Analysis of the activity of chains of internuncial neurons. *J. Neurophysiol.* **1**, 2–7.

15 J. W. S. Pringle, 1951. On the parallel between learning and evolution. *Behaviour*, **3**, 174–315.

16 K. J. W. Craik, 1952, *The Nature of Explanation*, Cambridge University Press.
17 C. Darwin, 1871, *The Descent of Man*, London, John Murray.
18 H. B. Thompson, 1899. The total number of functional cells in the cerebral cortex of man, and the percentage of the total volume of the cortex composed of nerve cell bodies calculated from Karl Hammarbergs's data; together with a comparison of the number of giant cells with the number of pyramidal fibers. *J. comp. Neurol.* **9**, 113–40.
S. P. Thompson, 1910, *The life of William Thompson*, London, Macmillan, II, 830.
19 J. Z. Young, 1963. The number and sizes of nerve-cells in Octopus. *Proc. Zool. Soc. Lond.* **40**, 229–54.
20 F. Dujardin, 1850. Sur le Système nerveux des insectes. *Ann. Sc. nat. Zool.* (3rd series), **14**, 195–205.
21 M. Rockstein, 1950. The relation of cholinesterase activity to change in cell number with age in the brain of the adult worker honeybee. *J. cell. comp. Physiol.* **35**, 11–23.
22 E. D. Adrian, 1947, *The Physical Background of Perception*, Oxford, Clarendon Press.
23 J. C. Eccles, 1953, *The Neurophysiological Basis of Mind*, Oxford, Clarendon Press.
24 C. F. A. Pantin, 1954. The recognition of species. *Sci. Prog., Lond.* **43**, no. 168, pp. 578–98.
25 G. A. Horridge, 1957. The co-ordination of the protective retraction of coral polyps. *Phil. Trans.* B, **240**, 495–529.
26 R. K. Josephson, R. F. Reiss & R. M. Worthy, 1961. A stimulation study of a diffuse conducting system based on coelenterate nerve nets. *J. Theoret. Biol.* **1**, 460–87.
27 C. F. A. Pantin, 1935. The nerve net of the Actinozoa. I. Facilitation. *J. exp. Biol.* **12**, 119–38.
28 C. F. A. Pantin, 1935. The nerve net of the Actinozoa. III. Polarity and after-discharge. *J. exp. Biol.* **12**, 156–64.
29 E. J. Batham & C. F. A. Pantin, 1950. Phases of activity in the sea-anemone, *Metridium senile* (L.), and their relation to external stimuli. *J. exp. Biol.* **27**, 377–99.
30 E. A. Robson, 1961. The swimming response and its pacemaker system in the anemone *Stomphia coccinea*. *J. exp. Biol.* **38**, 685–94,
31 E. A. Robson, 1963. The nerve-net of a swimming anemone, *Stomphia coccinea*. *Quart. J. micr. Sci.*, **104**, 535–49.

Notes and references, appendix 3

Appendix 3: pp. 171–184

1 C. Darwin, 1859, *The Origin of Species*, London, John Murray, p. 80.
2 L. J. Henderson, 1913, *The Fitness of the Environment*, New York, Macmillan, p. 5.
3 M. Faraday, 1844. A Speculation Touching Electric Conduction and the Nature of Matter, *Phil. Mag.* 3rd ser., **24**, 136–44.
4 C. F. A. Pantin, 1963, 'The Ballard Mathews Lectures', in *Science and Education*, Cardiff, University of Wales Press, pp. 1–53, p. 12.
5 H. H. Price, 1932, *Perception*, London, Methuen, p. 1.
6 A. Eddington, 1952, *The Expanding Universe*, Cambridge University Press.
7 W. Heisenberg, 1958, *The Physicist's Conception of Nature*, London, Hutchinson.
8 M. Faraday, *loc. cit.* p. 141.
9 See 7 above.
10 K. J. W. Craik, 1952, *The Nature of Explanation*, Cambridge University Press, p. 3.
11 C. F. A. Pantin, 1954. The recognition of species. *Sci. Prog. Lond.* **43**, no. 168, pp. 578–98.
12 W. H. Thorpe. 1950. A note on detour experiments with *Ammophila pubescens* Curt. *Behaviour*, **12**, pp. 257–63.
13 K. von Frisch, 1954, *The Dancing Bees*, London, Methuen.
14 D. Ilse, 1937. New observations on response to colours in egg-laying butterflies. *Nature, Lond.* **140**, pp. 544–5
15 R. J. Pumphrey, 1940. Hearing in insects. *Biol. Rev.* **15**, 107–32.
16 C. Darwin, 1871, *The Descent of Man*, London, John Murray, **1**, 145.
17 K. von Frisch, 1954, *The Dancing Bees*, London, Methuen.
18 C. F. A. Pantin, 1965. *Learning, world-models and pre-adaptation. Animal Behaviour Supplement* no. 1, pp. 1–8, London, Baillière, Tindall and Cox.
19 H. H. Price, 1932, *Perception*, London, Methuen.
20 H. H. Price, *op. cit.*
21 F. R. Winton and L. E. Bayliss, 1948, *Human Physiology*, London, J. & A. Churchill.
22 C. S. Sherrington, 1906, *The Integrative Action of the Nervous System*, New Haven, Yale University Press.

Index

Index

Index

Index

Index